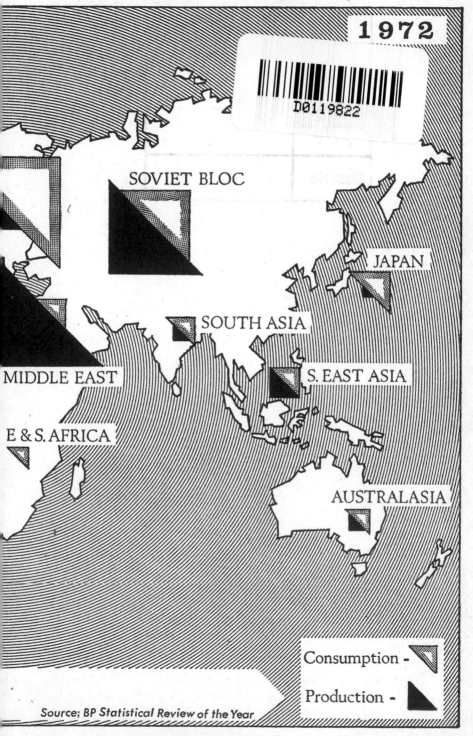

Overseas Gp 10.

1972

D0119822

SOVIET BLOC

JAPAN

SOUTH ASIA

MIDDLE EAST

S. EAST ASIA

E & S. AFRICA

AUSTRALASIA

Consumption -

Production -

Source; BP Statistical Review of the Year

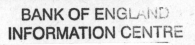

Oil
the biggest business

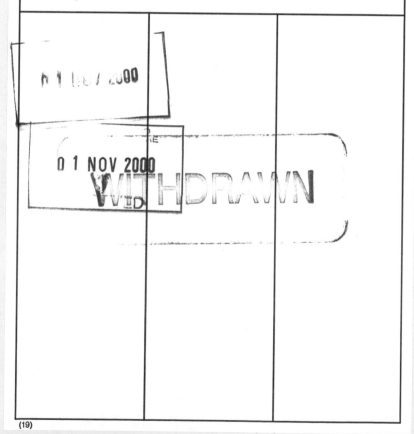

Oil
the biggest business

Christopher Tugendhat
and
Adrian Hamilton

Eyre Methuen
London

First published in 1968
by Eyre & Spottiswoode (Publishers) Ltd
New and revised edition 1975
© 1968 *Christopher Tugendhat*
© 1975 *Revised edition,*
Adrian Hamilton and Christopher Tugendhat
Printed in Great Britain for
Eyre Methuen Ltd
11 *New Fetter Lane, London EC4P 4EE*
by Cox & Wyman Ltd,
London, Fakenham and Reading

ISBN 0 413 33290 X

Contents

Plates

Acknowledgements and thanks for permission to reproduce photographs are due to Shell Photographic Service for plates 1, 3b, 3c, 4a, 4b, 5, 7, 8, 10a, 10b and 15; Esso Petroleum for plates 2 and 13; Continental Oil for plate 6a; Associated Press for plates 6b and 11; Camera Press for plate 6c; Press Association for plate 9; British Petroleum for plate 12; and Kuwait Oil Company for plates 14 and 16.

Figures

Figures 1–4 were drawn by Sidney Perrin. The rest of the figures, including the endpapers, were drawn by L. P. Thomas, except for Figure 21, which is reproduced from *Oil for the World* by Stewart Schackne and N. D'Arcy Drake (Harper and Brothers, New York).

Tables

Preface to the First Edition

This is an outsider's view of the international oil industry. I have been writing about oil for *The Financial Times* for several years. When I began I found that there was no independent and objective account of how the industry grew up, and how it works today. So I determined to try to write one.

This book is the result. It has not been sponsored in any way by any government, company, or other interest, and all the conclusions are my own.

CHRISTOPHER TUGENDHAT
London, October 1967

Preface to the Second Edition

Since this book was first published in 1968 the oil industry has been transformed. Part I, covering its history up to the formation of the Organization of Petroleum Exporting Countries (OPEC) in 1960, has been left untouched. But Part II has been entirely re-written by Adrian Hamilton. He started from scratch, and I have made no attempt to influence either his views or analysis. The result is therefore a joint effort. I am grateful to him for taking my foundation as the basis on which to build his edifice. His explanation of how the oil industry works, and his assessment of the energy situation, provide an invaluable insight into our present problems and a perceptive guide to the future.

CHRISTOPHER TUGENDHAT
London, March 1974

Acknowledgements

Among those who have helped me in my quest for information and knowledge my father, Dr Georg Tugendhat, holds a special position. After a lifetime in the industry he is omniscient about its affairs. I also owe a substantial debt of gratitude to Dr Paul Frankel and Mr Walter Newton, who have given me generously of their time.

I have spoken and argued with so many people in so many companies and governments that it would be invidious for me to single out any by name. But this does not diminish my gratitude, and I would like to take this opportunity to thank British Petroleum, Shell, and Standard Oil (New Jersey) for supplying the figures and other information for the statistical tables and diagrams.

Throughout the time I was writing this book I was employed by *The Financial Times*. I am deeply grateful to the Editor, Sir Gordon Newton, for his kindness and consideration. I joined the paper straight from university. He turned me into a journalist, and without his training I could never have embarked on this task.

The burden of my research work has been made much lighter by the help of *The Financial Times* Library staff, particularly the librarian, Mr W. Kuhnberg, and Mr Brian Carter. I am also grateful to Mr Sidney Perrin of *The Financial Times* for drawing the illustrations. Mrs P. Wadsworth of the Griffin Literary Service and Miss Fredy Dennis typed the manuscript, and without their assistance I would not have delivered it on time.

I should like to say that without Professor Allan Nevins' biography of John D. Rockefeller, *Study in Power*, and the late Robert Henriques' biography of Marcus Samuel I should never have been able to master the details of the industry's early history.

Finally, I have discussed many of the ideas incorporated in this book with John Eidinow and he has also read the manuscript. His help has been invaluable.

While writing this book I met and married my wife. She has been more understanding and helpful than any husband has a right to expect.

C.T.

ACKNOWLEDGEMENTS

To these acknowledgements should be added, for the second part of the book, my sincerest gratitude to George Glass and John Drummond of Shell, Graham Sterry of BP, Andy Campbell of Esso Petroleum, Andy Ensor of Mobil, Frank Waddams of the Polytechnic of Central London, Richard Johns and William Keegan of *The Financial Times* and the many, many others in the oil industry, government and institutions alike who have given so patiently of themselves, be their time ever so pressed with the dramas of the moment. If large corporations, private or public, have a monopoly it is the monopoly of information. It is not a power that they always use either wisely or well. But, when the oil companies do give, they give with a rare generosity and understanding, of which I have always been the recipient – even though these chapters may show just how much I have still to learn.

A. H.

CHAPTER 1

Introduction

In the Middle Ages men searched for the elixir of life so that they might have the gift of everlasting youth, and for the philosopher's stone, which would turn base metals into gold. Not even modern medical science can offer immortality, but in oil mankind has found a substance that is even more valuable and versatile than the philosopher's stone.

Oil means money. The Rockefellers, the Gulbenkians, and the Gettys owe their fortunes to it, and so do many other of the world's richest families; even the Rothschilds were once oilmen in the industry's turbulent early days before the First World War. Oil still retains its Midas touch, but today the beneficiaries are usually countries and their rulers rather than individuals. Kuwait, Abu Dhabi, and the Saudi Arabian royal family have taken over the aura of unlimited wealth that once used to belong to the Indian maharajahs.

Our civilization depends on oil more than any other single commodity. There seems to be almost nothing that it cannot do. In the nineteenth century, in the days of John D. Rockefeller, it was used as kerosene to provide light before the development of electricity. With the arrival of the internal-combustion engine it became the fuel for land, sea, and air transport; now it is being used in space rockets. It is the world's principal lubricant, and in industry it is the main provider of heat and power.

Yet to burn oil is incredibly wasteful; it could be even more useful as a chemical raw material. Already it is the basis from which such diverse products as plastics, detergents, and nylon, terylene, and other synthetic fibres are made. Even fertilizers and weed-killers derive from it; and perhaps within a few years, once scientists have succeeded in transforming laboratory experiments into full-scale production lines, it will become a major source of protein.

Natural gas, which is usually found in conjunction with oil and produced by the same companies, is only slightly less useful; it cannot be used as petrol or as a lubricant, but otherwise it is virtually interchangeable with oil, and often more efficient.

A country which is entirely cut off from oil cannot survive. In time of peace an effective embargo is almost impossible to impose, as the Rhodesians have shown. But war is a very different matter. Between 1939 and 1945 one of the main weaknesses of the German and Japanese economies was their lack of oil. This played an important part in their decisions to extend the area of conflict, and in the end was a crucial factor contributing to their defeat.

Since the war oil has become the most important commodity in international commerce. By an accident of fate most of the large fields are situated in the Middle East, North Africa, Latin America, and other places thousands of miles away from the main areas of consumption. Enormous fleets are needed to take it to Europe, North America, Japan, and the other industrialized countries, and oil now accounts for more than half the world's seaborne trade. There are more tankers in service than any other sort of ship.

In general, the producing areas have virtually no other natural resources. They are usually either deserts, mountains, or jungle, and without oil their people would be among the poorest and most backward. As it is, the countries concerned now receive vast annual revenues, and can threaten the prosperity of countries far more powerful than themselves simply by cutting off their exports. This is, of course, a double-edged weapon; nevertheless, it gives their governments more influence in international affairs than any comparable group of nations.

The oil industry itself is bigger and more international than any other. Its operations span the globe, and the companies conduct their affairs across national frontiers without reference to political differences. They negotiate with governments on almost equal terms, signing agreements almost as if they were sovereign independent states, and their finances dwarf the national budgets of all but the largest countries. Leaving aside the Communist bloc,

FIG I. SOME PRODUCTS DERIVED FROM OIL

Primary derivatives	*Typical end uses*
Carbon black	Tyres, plastics
Synthesis gas	Explosives, fertilizers, wood adhesives, animal feeds, nitrogen fertilizers, paint resins, anti-freeze, synthetic fibres (polyesters)
Methane	Shower curtains, toys, pipe, adhesives, plastics, synthetic rubber, solvent, refrigerant, fire extinguisher, solvent, pharmaceuticals, rayon, fumigants
Ethane	Toiletries, acetic acid, anti-freeze, detergents, paints, plastics, packaging, housewares, toys, tubing, polyvinyl chloride (see 'Methane' above), foam insulation, packaging
Propane	Gasketing, seals, solvent, epoxy resins, detergents, insulation, paints, plastics, synthetic rubber, synthetic fibre (acrylics), housewares, auto interiors
Butane	Nylon, plastics, synthetic rubber, lacquer solvents, tyres, windows, oil additives, adhesive
Benzene	Resins, pharmaceuticals, nylon, detergents, boats, auto parts, dyestuffs, synthetic rubber, insecticides
Toluene	Nylon, explosives, paint solvents
Xylenes	Paint solvents, paint resins, reinforced plastics, phthalic anhydride, fibres (polyesters)
Naphthalene	Dyestuffs, insecticides

the estimated cost of their property, plant, and equipment is over £50,000m., and nearly as much again will have to be spent in the next ten years to meet the expected increase in demand.

Despite its size and scope, the international oil industry is dominated by a small group of companies. Altogether there are seven, usually referred to generically as the majors, which together own well over a third of the existing investment, and account for more than half the sales. Five of these are American – Standard Oil (New Jersey), Texaco, Gulf, Mobil, and Standard Oil of California; then there is British Petroleum, which is almost half owned by the British Government, and Shell, which is Anglo-Dutch with twin headquarters in London and The Hague. These companies have the reputation both for indulging in desperate, unfettered competition and for forming close-knit international cartels. There is an element of truth in both views, although neither quite corresponds with reality.

The seven major companies, 1973

		Average net assets $m.	Net income $m.	Rate of return on assets %
Exxon	US	12,993	2,443	18·8
Royal Dutch Shell	Dutch/British	9,852	1,780	17·3
Texaco	US	7,583	1,292	17·0
Gulf	US	5,489	800	14·6
Mobil	US	5,430	849	15·6
Standard Oil of California	US	5,513	844	15·3
British Petroleum	British	4,439	760	13·8

The aim of this book is to explain how the industry works. It will show who does what, to whom, and when; it will describe how the companies conduct their operations, how they compete and how they don't. It will deal with the way in which their relations with governments in the producer and consumer countries are developing. It will also show how the range of uses for oil and natural gas is still expanding, so that in future they will play an even more important role in the world's economy than at present.

In many respects the industry's activities defy rational explanation. They are the result of its history, and events which took place many years ago still influence the way things are done today. To describe a nation's political system would be impossible without discussing how it grew up, and the same applies to the international oil industry. Thus the first part of this book is devoted to the period from 1859, when the first well was drilled, to 1960, when the governments of the producer nations formed the Organization of Petroleum Exporting Countries, usually known as OPEC. The Middle East crisis of 1973–4 can already be seen as another watershed. However, a final assessment of its significance will have to wait for several years. At this stage it is impossible to do more than discuss its immediate implications and to speculate on their future effect.

The second part of the book takes these developments on through the climactic crises in supply and price of the early seventies and examines just what impact these revolutionary developments will have on the future shape of the oil industry and energy scene. In the complete re-evaluation of the future that has followed in the wake of these changes, almost every issue is up for debate from whether the world really does face a fundamental crisis in energy supplies or whether the oil industry as we know it will survive at all. Figures of supply and demand and price have all had to be revised in an atmosphere of uncertainty over the future. Relations between the consumers and producers have been looked at without any real confidence in how they might develop. The place of the international oil companies within the structure of the industry has come under close scrutiny as both the producers and consumers have considered ways of bringing it under closer control and perhaps by-passing it altogether.

At this stage, answers to these questions are almost impossible to give with any degree of assurance. All that can be done is to define the issues and suggest the possible lines on which they may develop in what remains a highly volatile situation. But predictable or not, the questions themselves are of crucial importance to a world in which oil has become the fundamental fuel of industrial growth and in which the oil industry continues to be the biggest and most singular industry of our age.

PART I

How the Industry grew up
1859–1960

FRONTPIECE

Oil!
Beneficent oil,
Mankind's most precious treasure in the soil!
Oil!
Disgusting oil,
Father of blood and sweat and tears and toil!
Oil, you have made this puny race
Masters of time and Lords of space,
Have opened vast horizons for the poor,
And brought the city to the cottage door,
Or rather (which is not so good)
The cottage door to Hollywood.
Oil, you have made the mountains and the seas
Mean less than barbed wire fences mean to bees.

Methinks I see this writing in the sky:
'Those who by oil have lived by oil shall die'.

SIR ALAN HERBERT

Published in Punch *27 August 1941*
And dedicated to Georg Tugendhat and Franz Kind
Reproduced by permission of the author

CHAPTER 2

How it all began

Man has known about oil since time immemorial. The Babylonians used it as mortar, the Byzantines as Greek fire, and the Red Indians as war paint. In the eighteenth century the French began using it as a lubricant, and by the middle of the nineteenth oil lamps were lighting the streets of Bucharest. Scientists everywhere realized that oil – or petroleum, as it was often called – was potentially a very valuable substance, but nobody was able to produce it in large enough quantities to justify commercial exploitation; they either dug pits where it seeped through the earth's crust or skimmed it off the surface of streams. The problem remained unsolved until 1859, when almost simultaneously in the United States, Canada, and Germany efforts were made to drill for oil as if it were water. The only pioneer to be remembered is Edwin L. Drake, whose well near Titusville in Pennsylvania led to the discovery of the world's first major field.

He was working on behalf of George H. Bissell, a young lawyer from New York, who had become interested in the possibilities of oil during a visit to his old college at Dartmouth during the winter of 1854. The professors there had shown him a sample of crude oil with which they were experimenting in the laboratories, and told him that if suitably refined it could provide a better lamp light than kerosene produced from coal. At the time this was an astonishing claim to make; the process of manufacturing 'coal oil' had been patented in Scotland only four years previously, and with the growing shortage of whale oil and candle wax, which had hitherto been the main sources of light, it was believed to have a great future.

Bissell was so impressed that he promptly formed a company to buy the farm near Titusville where the sample had been found.

He also asked Benjamin Silliman, Jr, the professor of chemistry at Yale, for a second opinion. Silliman said that the Dartmouth scientists were right, and soon afterwards some of the more enterprising coal oil dealers confirmed this view by offering $20 a barrel* for all the crude oil they were offered. The only obstacle apparently standing between Bissell and a fortune was his inability to produce his oil on a substantial scale, and the company was losing money badly.

It was at this point that Drake appeared on the scene asking for a job and offering to put up some capital. He was hardly a man to inspire confidence. He had virtually no formal education, and had spent his life moving from job to job. Although he liked to call himself Colonel, he had never actually been in the Army, and the only uniform he had ever worn was that of a railroad conductor. But since he had some money, the company was prepared to accept him, and he was sent to Titusville to take charge of production.

Drake knew how water wells were drilled, and decided that as the existing methods of producing oil were so inefficient, drilling might be more successful. When he set up his derrick in the summer of 1859 the locals christened it 'Drake's Folly'. But a few weeks later on 27 August the folly paid off when oil was struck at a depth of 69½ feet. There were no dramatic gushers or explosions of the sort that were later to occur in Texas. The oil just rose to the surface under its own momentum, and when a pump was installed the well began producing at the rate of 30 barrels a day.

At the prevailing prices this meant a daily income of $600, and there was no chance of keeping Drake's discovery a secret. Within twenty-four hours hundreds of people were milling round Drake's well to find out how it worked, and prospectors and speculators poured into Titusville. The place became like a gold-rush town, with leases on promising land changing hands for thousands of dollars, and fortunes being made overnight. William Barnsdall, a tanner from England, who drilled the second well, cleared $16,000

---

* 1 barrel = 35 imperial gallons, or 42 American gallons. There is an appendix at the end of the book giving all the conversion factors normally used in the oil industry.

in five months, and a well called Mapleshade eventually made
$1·5m. for its owner, a local storekeeper called Charles Hyde.

The demand for the new product was insatiable, and by the
end of 1861 kerosene made from oil had entirely displaced the coal
variety. It also began to make an important contribution to the
war effort of the Northern States in the Civil War, since they
desperately needed a new source of foreign exchange to compensate for the loss of the South's cotton.

FIG. 2

Exports began in dramatic circumstances in December 1861
when the brig *Elizabeth Watts* sailed from Philadelphia to London
loaded with barrels of kerosene. While the cargo was being
brought on board the crew found out what they were handling
and promptly deserted for fear of being burned alive, so that at
the last minute the owners had to recruit another from among the
drunkards in the bars along the Delaware River. The voyage,
however, was a success; by the end of 1865 Britain, France, and
Germany were all substantial buyers, and for most of the rest of
the century exports accounted for over a third of the United
States annual production.

Sales rose rapidly, but after the first few halcyon months fortunes in oil were being lost as well as made. Unlike gold, oil does
not have an intrinsic value, and when production outruns demand

prices are bound to fall. The speculators overlooked this vital fact when they paid vast sums for a piece of land and then tried to pump oil out of the ground as fast as possible in order to recoup their investment. During 1860 prices dropped from the original $20 a barrel to $2, and during the war they fluctuated wildly from 10 cents a barrel, which made the wooden barrels more valuable than the oil they contained, up to $14.

Many of the original companies were forced out of business, and among them was Bissell's, leaving Drake again without a job. For a while he managed to earn a profitable living as Titusville's justice of the peace; that did not last for long, and when he died in 1880 he was a virtual pauper.

The day of the small man was coming to an end. In the years after the Civil War it soon became apparent that the industry's two great weaknesses were its lack of capital and of large companies. In the oil regions of Pennsylvania production was in the hands of dozens of tiny concerns, many of them one-man affairs, and the same was true of the refineries, which were springing up all over the country. Nobody was strong enough either to control the flood when a prolific new well was discovered or to build up sufficient stocks to maintain supplies during a temporary shortage. The time was ripe for consolidation, and the man who saw the opportunity most clearly was John D. Rockefeller.

CHAPTER 3

The rise of Rockefeller

'Oil warfare, in the latter part of the last century and the first decades of this, was not an activity which can be judged today, by ordinary people living ordinary lives, according to any accepted civilized standard.'

Robert Henriques

Rockefeller's career before 1865 was in the classical nineteenth-century American tradition of self-help, Godliness, and private enterprise.

He was born on 8 July 1839 on a farm near the village of Moravia in western New York, and when his parents moved west to Cleveland he went with them. His father was something of a rogue; in Moravia he was indicted for rape, though never tried, and later held a variety of rather dubious jobs, at one time advertising himself as 'Dr William A. Rockefeller, the Celebrated Cancer Specialist'. He was often away from home, and John, as the eldest of six children, had to help his mother raise the family. Like many other ill-used wives, she was a religious woman, and the boy took after her. In 1854 he was accepted into full membership of the Erie Street Baptist Church in Cleveland, and for the rest of his life religion and church affairs remained one of his principal interests. Even as a boy, his enthusiasm was so great that as soon as he was twenty-one he was elected one of the five trustees of the Erie Street community.

Rockefeller's religion was never simply a 'Sundays only' affair to be forgotten as soon as the sermon was over. From the moment he started earning money of his own he began giving it away. He always kept records, which reveal that in his first four months at work his total pay amounted to less than $95; out of this he gave away $9 09. The recipients seem to have been whoever needed help most; many of the entries simply read 'poor man' or 'poor woman', and he never seems to have recognized religious or racial distinctions. In 1859 he helped a Negro slave to buy his wife, and during the Civil War he contributed to a Roman Catholic orphanage as well as a Swedish mission, which shows incredible broadmindedness for so bigoted an age.

From the very first he seems to have been ambitious: 'I did not guess what it would be, but I was after something big,' he explained many years later.* He refused to take any job that would not give him a training to establish a business of his own, and in 1855 he managed to become a book-keeper in a firm of commission agents and produce shippers. Three years later, when his employer refused to give him a large enough salary, Rockefeller felt confident enough to strike out on his own. In 1859, with some financial help from his father, he went into partnership with a young emigrant from England called Maurice Clark to establish the commission agency of Clark and Rockefeller.

With the outbreak of civil war they could not have chosen a better line of business. The rapid mobilization of the Northern armies created an insatiable demand for foodstuffs and clothing of every kind, and as prices rose, so the percentages going to the commission agents increased in value. Clark and Rockefeller dealt in everything they could lay their hands on, and within a short time this naturally included kerosene.

In the early 1860s Cleveland was particularly well placed to become a refining centre. It had a direct rail link with the oil regions in Pennsylvania, and unrivalled communications with the East through two railways and the Great Lakes with their interlocking system of canals. Several refineries were soon operating in the city, in one of which worked Samuel Andrews, another English emigrant from the same town in Wiltshire as Maurice Clark. Andrews enjoyed a reputation for being able to extract more kerosene from a barrel of crude oil than anybody else, and in 1863 Clark and Rockefeller agreed to form a refining company in partnership with him.

The venture prospered, and by 1865 the refinery was the biggest

---

* *Study in Power, John D. Rockefeller, Industrialist and Philanthropist,* Allan Nevins. Charles Scribner's Sons.

in Cleveland. It was also expensive to run. The handling of large quantities of kerosene required much greater capital investment in plant, storage facilities, and distribution than the rest of the commission agency. Rockefeller was prepared to borrow as much money as was needed, but Clark was doubtful, and relations between them deteriorated. In January 1865 matters came to a head. They decided to part, and on 2 February Rockefeller paid $72,500 and handed over his share in the commission agency for Clark's share of the oil business.

$72,500 was a large sum for a young man of 26, and Rockefeller said later that he had to borrow so heavily that, 'I had worn out the knees of my pants'* begging credit from the banks. But this was an exaggeration. In a business regarded as notoriously unreliable, his previous successes as a commission agent and his position in the Baptist church gave him an unusual standing. The bankers realized that if only sound men would go into oil, refining had a great future, and they were pleased to back a man with Rockefeller's credentials.

Nevertheless, a new partner was needed who could bring more money to the business, and in 1867 Rockefeller invited Henry M. Flagler, a successful grain merchant, to join Andrews and himself in the refinery. Flagler's father-in-law, Stephen V. Harkness, was a notable entrepreneur who had made a fortune out of the war by buying up large stocks of whisky and selling them at a profit after the imposition of the special war taxes. When Flagler accepted Rockefeller's offer Harkness contributed between $60,000 and $90,000, while remaining a sleeping partner. We don't know how much Flagler himself invested, but his main contribution was undoubtedly his business acumen. Rockefeller had complete trust in him, and relied heavily on his new partner's judgement. It was Flagler who carried out the vital negotiations with the railroad officials over freight rates when the business started to expand, and it was also he who originally suggested that the firm should become a joint-stock company.

As a joint-stock company it would be easier to raise money and to take over other companies by offering their owners shares. The proposal was accepted, and on 10 January 1870 the Standard

* Nevins, op. cit.

Oil Company was incorporated in Cleveland with Rockefeller as president, William Rockefeller, his younger brother, as vice-president, and Flagler as secretary and treasurer. Andrews held no official position in the corporation structure, but received a large block of shares and remained in charge of refining. There is no doubt that from the outset the partners intended that Standard should take over as many of its rivals as possible. In the following year the company's capital was increased from $1m. to $2·5m. expressly for this purpose, and in their correspondence Rockefeller and Flagler referred to their proposed expansion as 'our plan'.

Rockefeller realized that the possibilities of oil were limitless so long as Standard expanded in the right direction, and he was acute enough to see that in the long run this meant concentrating on refining. He was not dazzled by the enormous fortunes that could be made overnight in the fields in the Pennsylvania oil regions by men like Barnsdall and Hyde. He understood that exploration and production, like any other form of prospecting, was simply a gamble in which far more men lost than won. He saw, too, that a production business could not be run on sound commercial lines. There was always the danger of a new well coming onstream near by and knocking the bottom out of the market; or, in those early days, a man's well might suddenly dry up for no apparent reason. Refining, by contrast, could be run like any other manufacturing undertaking, and it had the additional advantage of providing the vital link between the producer and his final market. It was the Gibraltar of the industry, and the man who controlled it could dominate the producer and retailer.

But it was no use simply being one among many units. Rockefeller was convinced that profitability depended on consolidating nearly all the country's refineries – in New York, Pittsburgh, the regions and elsewhere, as well as in Cleveland – into one great organization. This would then be powerful enough to stabilize both the prices it paid for its crude oil and those it received for the refined kerosene from the retailers and distributors. While the industry was made up of hundreds of small units there was no possibility of earning a consistent profit. On the one hand, the price of crude oil fluctuated wildly, depending on the level of production in the regions and the number of new wells that were being drilled, while, on the other hand, refiners had to carry on cut-throat competition with each other to win contracts from the retailers. Rockefeller estimated that in 1871 'more than three-fourths of the oil refiners of the country did a losing business',* and he saw no hope of the situation improving until they came together.

Another advantage for a large consolidated company would be that it could secure favourable terms from the railroads. In the years after the Civil War American industry and farming were dominated by the big rail networks. They could make a township by linking it to their systems and so connecting it with the outside world, or ruin it by going round to another. Many communities in the West were destroyed in this way when they refused to pay the railmen's prices. Industrial concerns were equally vulnerable, and by manipulating their freight rates the railroads could threaten a company with bankruptcy.

The only way to be sure of securing favourable rates was to become the largest unit in a particular industry. Despite their power, the railroads were never able to organize any lasting market-sharing agreements, and wherever their lines were in competition with each other they engaged in rate-cutting wars in an effort to secure the largest volume of freight. Consequently, the company, or group of companies, able to guarantee this could invariably secure cheaper transportation than its rivals. Cleveland was fortunate in being served by two lines to the East, and by the late 1860s it also had a choice of several into the regions. Rockefeller was determined to become by far the largest shipper of oil, so that they would want to compete for his traffic. In 1870 Standard began taking over some of the smaller Cleveland concerns, and by the end of 1871 it was probably the largest single refining company in the country. But before it reached a position to exert any real leverage over the industry the leading railroads took the initiative out of his hands.

The three main oil-carrying lines, the Pennsylvania, the Erie, and the New York Central, came forward with a plan of their own for simultaneously stamping out competition in both the rail and

* Rockefeller papers.

oil industries. This was to be achieved by forming the thirteen largest refining companies and the three rail carriers into two associations. Each would then grant special terms to the other, and the scale of their operations would be such that the outsiders would not be able to compete. The oilmen undertook to form a joint transportation concern, with the rather quaint name of the South Improvement Company, and to send all their shipments along the rail association's tracks with a specified quota for each, thereby putting an end to competitive freight rates. For their part the railmen promised to give the South Improvement Company much lower rates than anybody else. Rockefeller was doubtful that a scheme involving so many different interests could succeed, but he was willing to give it a try; competition among the refiners had reduced prices to rock-bottom levels, and if the joint venture failed it would afterwards be easier to persuade the other refiners to throw in their lot with Standard. In any case he was not yet strong enough to stand out on his own against the railways and the other large refiners acting together.

The precise terms of the arrangement were drawn up at a meeting on 2 January 1872, when the quotas for the three rail-roads were worked out. At the same time the refiners agreed that the railroads should substantially increase the published rate for oil shipments, and on some routes these were doubled. In return the railroads agreed to give the members of the South Improve-ment Company large enough discounts to offset most of the in-crease on their own shipments and the same discount on every barrel of oil dispatched by their competitors. Thus the outsiders were not only to be faced with higher rates but were also to subsidize the activities of the South Improvement Company; and the more oil an outsider shipped, the larger this subsidy would be.

If the scheme had been put into practice it would have ruined the refineries outside the South Improvement Company within a matter of months, and left the crude-oil producers in the regions in a hopeless bargaining position when selling to the big refiners. But while the preparations for the *coup* were still being made the secret leaked out. The details of the new rate schedule had been sent to a freight agent in the regions with instructions to keep them secret until he received a final order from head office. The man had a son dying at home, and left his office in charge of his subordinates without passing the message on to them.* They published the new rates on their own responsibility, and in the morning papers of 26 February 1872 oilmen everywhere found out what was in store for them.

They exploded with wrath; a mass meeting was called in the Titusville opera house; a Petroleum Producers' Union was formed pledged to cut off supplies to 'the conspirators'; and demands for an investigation were sent to Congress and the Pennsylvania legislature. Within a few days a petition 93 feet long had been drawn up demanding that a Bill should be passed to enable a pipeline to be built that would break the railroads' monopoly, and an army of 1,000 men was kept ready to march on the state capital of Harrisburg.

The embargo on sales was effective, and within a short time the members of the South Improvement Company were forced to open negotiations with the union. There was also a Congressional Investigation in Washington at which they were forced to admit that the principal purpose of their scheme was to increase prices. Public opinion, understanding this from the first, was thoroughly roused against the project, and the participants were subjected to mounting criticism in the Press. Emotive phrases such as 'the black anaconda' were widely used, and the oil industry started to acquire its reputation for monopolistic tendencies, profiteering, and unfair trading practices.

The impression also grew up that Standard was responsible for the whole affair, which had somewhat less justification. The blame should have been put on the railroads, and particularly on Tom Scott of the Pennsylvania, whose idea it had been in the first place. However, after his initial doubts Rockefeller was all for standing firm – 'It may take weeks and months yet to succeed, but the Union of our American refineries is worth much labour and patient effort,' † he wrote to his wife early in March. The railroads by contrast were becoming increasingly nervous and anxious to jettison the entire scheme. When a committee from the producers'

* *The History of the Standard Oil Company*, Ida Tarbell. McClune, Phillips and Co., 1904.

† Nevins, op. cit.

union visited the formidable Commodore Vanderbilt of the New York Central he blamed the whole thing on his son Billy, and promised never to have anything to do with the South Improvement Company or any of its members. Tom Scott achieved an even more notable *volte face*; when the committee came to see him he agreed that the contract was unfair, and offered to sign a similar one with the producers instead. The committee declined, but one of its members was so impressed that on leaving the room he nominated Scott for the next president of the United States. Eventually at the end of March the producers and railroads composed their differences, and a new set of freight rates were drawn up. For a while the producers maintained their embargo, but by the end of April it had broken down, and the 'Oil War of 1872' was over.

Rockefeller returned to his own expansion plans, and within a few months the industry found that it was up against a far more determined consolidator than the South Improvement Company. When Standard began its take-over campaign it accounted for between 10 and 20 per cent of the total American refining capacity; by 1879 the proportion was somewhere around 90 per cent.

The most important single reason why Rockefeller was able to move so fast was that much of the rest of the industry was on the verge of bankruptcy. Although the demand for oil was still growing rapidly, there were far too many plants in operation, and many of them were working at only a fraction of their full potential. This had, of course, been the case ever since the end of the Civil War, and as the surplus increased, so profit margins declined; between 1865 and 1870 they dropped from 19·50 cents a barrel to 7·90. But men had been disinclined to leave the industry: most of them had sunk their entire capital into their companies, and they believed that one day prices must improve. In 1873 these hopes were finally dashed when the United States was hit by a general economic depression, almost as bad in its immediate effects as that of 1929. Thus when Rockefeller approached other refiners with an offer to take over their businesses most of them were delighted to accept. His price was usually much less than the amount of their original investment, but a sale was preferable

1. Drake's well. Edwin L. Drake is the bearded figure on the right

2. John D. Rockefeller

to continuing to operate at a loss, and there was no prospect of receiving a bid from anybody else.

Even the larger and more prosperous concerns were in a mood to listen to Rockefeller's suggestions. They, too, realized the need for the industry to close down its surplus plants, and they were attracted by the idea of a single organization large enough to dictate terms to the railroads, the producers, and the distributors. Some of them would no doubt have liked to be in a position to establish it themselves, but Rockefeller was clearly stronger than they were, and whenever he was dealing with a potentially dangerous rival he was sensible enough to hold out the prospect of an attractive position in Standard if only they would accept his offer. W. G. Warden of Warden, Frew & Company of Phila-delphia, Charles Lockhart of Pittsburgh, and Charles Pratt of New York all went straight on to the Standard board. As former associates in the South Improvement Company, these men were in a sense old friends, but Rockefeller was just as willing to be generous to his enemies if he regarded them as able or useful. Such a man was John D. Archbold, a leading figure in the regions, who had played a prominent role in organizing the opposition to the South Improvement scheme. He also was given a good job with Standard, and many years later he was to take over as chief executive from Rockefeller.

Nevertheless, it would be wrong to regard Rockefeller as being an angel of mercy for distressed oilmen. When somebody refused to sell he was quite prepared to use threats. In 1876 his younger brother Frank, who was in another company which tried to stand out against a take-over bid, told a Congressional committee that Rockefeller and Flagler warned that: 'If you don't sell, your property will be valueless, because we have advantages with the railroads.' * Frank was always jealous of his elder brother's success, and this may be an overstatement, but it certainly con-tains an element of truth. Because of its size, Standard could always secure favourable freight rates from the railroads, and rather than fall out with Rockefeller, the rail companies were prepared to make life difficult for other shippers.

There are several reasons why it should have been Rockefeller

* Nevins, op. cit.

who brought the industry together. At the outset he was undoubtedly considerably helped by the refining skill of Sam Andrews, which gave him a head start over his competitors. For a long time most refiners were interested only in producing the maximum amount of kerosene from their crude oil, and threw away the various by-products which had to be produced as part of the refining process. In Cleveland they were sometimes prosecuted for tipping their petrol into the river. But Andrews could find a use for everything: the petrol was burnt as fuel to heat the stills in the refinery, oil was sold to the gas plants to be used for gas-making instead of coal, and industry was persuaded to try lubricants made from oil instead of from vegetable oils. Without Andrews none of this might have been possible in the early years of Standard. But in 1878 he resigned on the grounds that the company had become too large, and as Standard continued to prosper, it cannot be said that his role was crucial to its success. Like most great men, Rockefeller knew how to choose his subordinates, and when they left him he was always able to find others just as efficient.

One reason for this was an incredible attention to detail which ensured that all his employees had to work hard. There are innumerable stories told to illustrate this, of which perhaps the most famous concerns the 'thirty-nine drops'. After watching solder being dropped on kerosene cans to seal them, Rockefeller asked the supervisor how many drops were needed. 'Forty', the man replied, at which Rockefeller asked if they had ever tried thirty-eight. When the answer was 'No' he suggested they should do so, and submit a report to him. This showed that with thirty-eight drops 6 or 7 per cent of the cans leaked, so another experiment was carried out with thirty-nine drops, and this proved to be sufficient. This particular economy cannot have meant much to Standard in financial terms, but Rockefeller's ability to find ways of cutting costs sometimes gave Standard important competitive advantages over its rivals. When, for instance, he realized that barrels and cans could be made more cheaply than his suppliers charged, Standard started to manufacture its own, with the result that its expenditure on these important items, without which oil could not be sold, was reduced to a level far below that of its rivals.

Another of Rockefeller's techniques for getting the best out of his subordinates was to encourage competition between them. Each refinery and plant under his control had to submit monthly reports on its activities, and these were compared. They were also made known throughout the organization, and used to inspire rivalry, in much the same way as a schoolmaster uses exams. Those who did badly had to adopt the techniques of those who did well, and consistent lack of success in a particular operation led to the men in charge being dismissed. There were generous prizes as well as dire punishments. It was Rockefeller who earned the oil industry the reputation for paying high salaries, which it still enjoys today, and executives were encouraged to buy shares of their own; if they could not afford to the money was lent to them on easy terms.

In the last resort, however, none of Rockefeller's management skills could have brought him success without his astonishing ability to raise money. To some extent his credit was derived from his record of business success, and his reputation for efficiency naturally inspired bankers with trust. But from the very beginning, when he borrowed enough to buy out Maurice Clark, he seems to have been possessed of an extraordinary knack for charming money out of bankers. Sometimes they would even approach him. Once he was walking along a Cleveland street with the president of the local YMCA, a Mr H. S. Davies, when a prominent local banker, Daniel Eells, stopped his carriage and in front of passers-by said: 'Mr Rockefeller, do you think you could use $50,000?' *

Not surprisingly, Rockefeller's success in consolidating the refining industry created mixed feelings among the railroads. They were glad to have a strong and reliable company to deal with that could guarantee large volumes of freight, but against this they soon started to fear the power Standard could exert. Rockefeller left the transportation side of the company's affairs to Flagler, who was highly skilled in playing one line off against the other and switching consignments from route to route in order to secure the most favourable rates. The railroads also began to wonder whether perhaps they might not be able to enjoy some of the profits which Rockefeller seemed to find in oil.

* Nevins, op. cit.

These hopes and fears were held most strongly by Colonel Joseph D. Potts, the head of the Empire Transportation Company, a freight-carrying subsidiary of Tom Scott's Pennsylvania. In the summer of 1873 the Empire took over two of the largest pipeline networks in the oil regions. In those days pipelines were rudimentary affairs laid along the surface, linking groups of wells with the railheads; they were not used to carry oil over long distances. Nevertheless, they were extremely important. If the Empire could establish a monopoly over connecting the wells with the railroads it could dictate the price which Standard would have to pay for oil for its refineries. Rockefeller immediately retaliated by sending his agents into the regions to take over as many of the pipeline companies as were still available, and for a while he was able to work in harmony with the Empire. Indeed, they ran a highly profitable market-sharing arrangement together. This broke down when Rockefeller discovered that the Empire's oil was being carried more cheaply than his own on the Pennsylvania's network. From that moment on Standard and the Empire were in competition with each other, and by 1877 it was clear that so far as Rockefeller and Potts were concerned it was war to the death.

The Empire carried its challenge into Standard's home territory by taking over refineries and building more of its own. Standard retaliated by cancelling all its contracts with the Pennsylvania, ordering a crash programme for the construction of new tank cars to carry oil on the rival railroads, and slashing prices wherever the Empire tried to sell oil. The two sides were equally matched, and Potts, with Scott's backing, was prepared for a long struggle. Then the railmen were struck by two unexpected blows: in April a bitter rate war broke out involving all the Eastern railroads, and covering passenger fares and freights; still worse, in midsummer the railroad workers came out on strike. It was one of the bloodiest labour battles of the nineteenth century, and in Pittsburgh the situation got completely out of control. The police opened fire on the strikers, killing twenty-five men, and looting and destruction took place on a massive scale. When peace was restored Scott found that his companies had lost more than 100 locomotives and 1,500 passenger and freight cars.

He had no alternative except to sue for peace. The result was a take-over of gigantic proportions, with Standard purchasing all the Empire's oil interests – refineries, pipelines, and tank cars – for $3·4m. in cash. A few weeks later Rockefeller purchased the largest remaining pipeline company, and by the end of 1877 hardly a barrel of oil could get to a railroad without his consent, whoever it belonged to.

Rockefeller was now more powerful than the South Improvement Company could ever have been: he not only controlled the oil industry, but also dominated the railroads, since it was only by dealing with him that they could get any worthwhile oil freight at all. The new relationship was formalized when he divided up his shipments between them, and they in turn undertook to give him discounts both on his own oil and on that dispatched by other companies.

The producers felt the noose of monopoly tightening around their necks, and hit out in all directions. They applied to the Governor of Pennsylvania to force the Standard pipelines to fulfil their legal obligations as a common carrier; they demanded action from the Pennsylvania State Supreme Court to prevent the railroads from discriminating in their freight rates; and they charged the Standard directors in person with criminal conspiracy. The case was withdrawn when the company agreed to reform its pipeline practices and the railroads made some more promises about rates, but nobody took these undertakings very seriously.

Yet the producers' actions did have one important effect; they led to official inquiries, at which the details of Standard's rebates were exposed. The American public was not easily shocked at that time by anything in business. It knew how the railroads adjusted their rates to help the big shippers and generally believed that in commercial life 'only the weak are good because they are not strong enough to be bad'. Nevertheless, when the inquiries revealed that the independent refiners and producers had to pay $1·44½ a barrel to send their oil from Titusville to New York, whereas Standard was charged only 80 cents, this was thought to be altogether too much. Rockefeller's reputation as an extortionate monopolist was firmly established in the public mind, and in later years it was to do him a great deal of harm.

In the short run he had more pressing problems to worry about. While the public outcry over the freight rates was at its height a small group of producers led by Byron D. Benson determined to launch a final desperate attack on Standard's monopoly by building a pipeline over the Allegheny mountains to the coast. In this way they could avoid the Standard pipelines and the railroads, and then ship their oil by sea, either to New York or to Europe. They formed themselves into the Tidewater Pipe Company, and hired General Herman Haupt, the most famous civil engineer in America, to take charge of the operation.

At first Standard and the railroads did not take the matter very seriously. Oil had never before been pumped in a pipe larger than 3 inches in diameter for more than 30 miles or over any significant hill. The Tidewater line was to be almost 110 miles long with a diameter of 6 inches, and it had to cross the mountains at a height of 2,600 feet. The experts said it could never be done. Rockefeller believed them, but none the less took the precaution of buying tracts of land across the proposed route. Even when Haupt found a way round most of these, it was generally believed that he could never complete the project. In the summer before the pipe was buried the heat made it writhe about like a giant snake, knocking down trees and bushes; in the winter all the materials had to be hauled for 30 miles through snow-covered passes on horse-drawn sleighs.

It seemed impossible that Haupt could ever triumph over such obstacles. Yet he did. On 28 May 1879 the oil began to flow, and overnight the industry's transportation was revolutionized. The Tidewater cut the cost of carrying a barrel of oil to the coast from about 40 cents to 17 cents, and though the railroads promptly cut their rates to 10 cents a barrel, it was obvious that their days as large-scale carriers of oil were numbered.

Rockefeller started building long-distance pipelines of his own, and while the work was in progress he set about trying to destroy the Tidewater. Benson and his friends wanted to refine oil themselves, and to sell to the independent refiners in order to have nothing to do with any of Standard's facilities. Standard foiled the plan by taking over yet more plants. It also tried the Trojan-horse technique of secretly buying Tidewater shares in an effort to gain control of its rival. When this failed it went a step further and attempted to throw the company into bankruptcy by bribing another stockholder, a Mr E. G. Patterson, $20,000 to ask the courts to appoint a receiver.* The case was dismissed, but by the autumn of 1883 the Tidewater had had enough. Standard by now owned extensive lines of its own, and the competition was becoming ruinously expensive. When, therefore, Rockefeller suggested one of those generous settlements which he customarily offered to opponents he respected, the Tidewater's leaders preferred to accept rather than face the much stiffer competition which they knew would follow a refusal.

The settlement gave them a guarantee of $11\frac{1}{2}$ per cent of the east-bound pipeline traffic, and the rates of the two companies were set at the highly profitable level of 40–45 cents a barrel, depending on the destination. Thus the Tidewater became a colony of Standard's empire, and Rockefeller's control over the industry became stronger than ever.

During the fight with the Tidewater Rockefeller carried out a major reorganization of Standard's internal affairs which was to have a profound effect on the whole economic life of the United States. He created the Standard Oil Trust, the first of the great trusts which enabled industrialists to overcome the strait-jacket of nineteenth-century American company law and to establish organizations that could operate across the nation without regard to state boundaries.

Standard's activities were, of course, already nation-wide. Yet it was legally impossible for the company, as an Ohio corporation, to own plants or shares in companies in other states. It overcame this difficulty by appointing one or other of its officers to take personal possession of the shares of the companies which it acquired. This officer would then hold the stock as trustee for Standard. The arrangement had the advantage that when Standard did not want anybody to know that it had acquired another firm the fact could easily be hidden. At the same time it hampered the creation of a unified management structure, which could not be achieved even by the establishment of a special board of trustees to hold the shares of the subsidiaries.

* Ida Tarbell, op. cit.

The man who found the answer to this problem was Samuel C. T. Dodd, the company's chief lawyer, who suggested creating a separate Standard company in each state to acquire all the group's assets in that state. There should be a common management for supervising all the subsidiaries, and their stock should be vested in a single board of trustees who, in exchange, would issue certificates of interest in the trust estate. On 2 January 1882 the necessary papers were signed, and a board of trustees headed by Rockefeller was established with headquarters in New York at 26 Broadway.

This was a ten-storey building on the site of a house which had once belonged to Alexander Hamilton, the champion of the concept of an industrialized America and the principal opponent of Jefferson's agrarian ideals. There is no reason to suppose that Rockefeller knew about this link with Hamilton, but it is nevertheless one of the finer coincidencies of American history. It is not often that the visionary and the man who later fulfils his dreams live on the same spot.

The trust did not last for long in its original form. In 1892 it had to be dissolved following a decision by the Ohio Supreme Court that it violated the original Standard charter, since most of the trustees were non-resident in the state. But by this time Dodd's plan had served its purpose. Some of the state governments were beginning to realize the advantages of adjusting their legal systems to commercial requirements, and New Jersey in particular had devised a liberal new company law enabling corporations to hold stock in each other. So Standard Oil (New Jersey) took over the trust's role as a holding company owning all the shares in the subsidiaries in the various states, and everything continued much as before.

To most of his contemporaries Rockefeller at this stage of his career was a colossus towering over the industrial world. First he had consolidated the refining industry, then he had dominated the railroads, and finally he had acquired a monopoly of the country's pipelines. His empire was welded together by a corporate structure that enabled him to control its activities in every state, and these in turn were crucial to the prosperity of the country. Nor was his power confined to the United States, for it was still the world's only large exporter of oil, and the greater part of this trade, too, belonged to Standard.

So far as Standard's profits were concerned, kerosene was still far and away its most important product, as factories, homes, and offices throughout the civilized world relied on it for their light. But some of the other products, which Sam Andrews had been so careful to make use of, were also becoming vital to the international economy. Lubricating oil, for instance, had entirely displaced vegetable lubricants in steam engines, machine tools, and locomotives, and fuel oil was beginning to be used in place of coal to heat the boilers in factories and ships. Oil was even being used on a small scale as a chemical raw material in the manufacture of paints, varnishes, dyes, and, most surprisingly of all, chewing gum.

Rockefeller was still very chary of moving into the production of oil. He continued to regard it as basically a mug's game, and preferred that other people should risk their money to find the oil his refineries consumed. Quite apart from this, he disliked everything about the oil regions, with their gambling houses, speculators, and drifters. It was an atmosphere in which he could never feel at home. But Standard could not afford to risk the possibility of running out of oil, and when new fields began to be discovered in Ohio and Indiana, and then scattered all over the continent, Standard decided to buy a few to run on its own account, although these operations were always on a modest scale.

The same cannot be said for the company's expansion into retailing and distribution. Once his consolidation of the refining and transportation sectors was complete, Rockefeller turned his attention to the distribution of his products, and after the formation of the trust he quickly acquired control over most of the smaller companies which had previously looked after this on his behalf. As a result, Standard was brought into direct contact with the general public, and the power and scale of its activities therefore became a cause of much wider concern than would otherwise have been the case.

In some respects the public gained from the change, since Standard was able to organize the distribution of oil far more efficiently than the companies it superseded. It was also willing

to go to almost any lengths to maintain the reputation of its product. Complaints from customers were dealt with promptly, and
in areas where the quality of the lamps and stoves was such that
the kerosene would not burn properly Standard was prepared to
undertake the supply of these articles as well. Foreign customers
were looked after just as carefully as those in the United States.
In 1883, when the West of England Petroleum Association cen-
sured the company for 'inexcusable carelessness', a senior officer,
F. W. Lockwood, was sent over to investigate the situation. He
reported back that the cause of the trouble was the poor quality of
English wicks, and the association was told that if the local manufacturers could not make the necessary improvements Standard
would provide an alternative source of supply. No other company in Britain or the United States could provide such service,
and there are not many that would do so today.

Nor could any other oil company in nineteenth-century
America grow large enough to challenge Standard. Rockefeller's
monopoly over the nation's pipeline system meant that nobody
else could transport oil over long distances without using his
facilities, and, despite its legal obligations as a common carrier,
Standard refused to allow this. On routes where railroads were
still used he again had a decisive advantage, since no railway
company could afford to alienate him, and he could usually insist on receiving a discount on his rival's shipments as well as his
own. This control over the transportation of oil meant that
Standard's refineries could invariably be sited in the best position
to supply particular markets, and that taking the country as a
whole, its operating costs were bound to be lower than those of any
other company.

Yet Standard never achieved a complete monopoly. Between
10 and 20 per cent of the refining capacity, and a similar proportion of the market for refined products, always remained outside
its control in the hands of a large number of small independent
concerns. These were to be found all over the country, and some of
them exported oil to Europe, but their individual activities were
inevitably on a small scale. They would draw their crude oil
supplies from a nearby oilfield, and so long as their sales of
kerosene and other products remained modest, Standard let them

live. To have tried to stamp out all opposition in a country as large as the United States would have been far too expensive to be worth the effort.

It was only when a company showed signs of wanting to capture valuable contracts from Standard or of expanding beyond a limited area that Standard 'cut to kill', as Miss Ida Tarbell put it in her famous history of the company. It was on these occasions that it employed the methods that have earned Rockefeller his reputation for extortion, rapacity, and disregard of the public interest.

The most commonly employed device was selective price cutting financed by over-charging in other places where Standard enjoyed a monopoly. This could lead to the most incredible fluctuations in price, and was also very unfair to the people and industries living in the high-price areas. In Denver, for instance, in 1892 Standard sold kerosene at 7½ cents a gallon in order to kill off some competitors, while elsewhere in Colorado the same oil was selling at 25 cents a gallon. This may have been an extreme example, but in the same year a survey in California showed that where Standard was meeting competition it sold oil at 17½ cents a gallon, whereas in those areas where it enjoyed a monopoly its price was 26½ cents a gallon.

As soon as the rivals had been disposed of, prices would rise again. Kerosene was usually sold through provision stores and other retail shops, and if a shopkeeper or dealer tried to buy from another supplier Standard was quite prepared to drive him out of business. 'The reason we quit taking your oil is this,' wrote a Kansas dealer to Scofield, Shurmer, and Teagle in 1896: 'The Standard Oil Company notified us that if we continued handling your oil they would cut the oil to 10 cents retail, and that we could not afford to do, and for that reason we are forced to take their oil or do business for nothing or at a loss.'* The following year an Ohio dealer wrote to the same firm: 'The Standard agent has repeatedly told me that if I continued buying oil and gasoline from your wagon they would have it retailed here for less than I could buy. I paid no attention to him, but yesterday their agent was here and asked me decidedly if I would continue buying oil

* Ida Tarbell, op. cit.

and gasoline from your wagon. I told him I would do so; then he went and made arrangements with the dealers that handle their oil and gasoline to retail it for 7 cents.'*

There was little chance of a dealer signing an agreement with an independent supplier that Standard did not know about if it was trying to bring an area to heel. Through bribery and corruption it ensured that it always knew exactly what was going on. The best source of information were the clerks at the rail freight depots, and these were frequently paid a regular retainer to pass on all details of other companies' oil shipments to the local Standard agent.

The company was also prepared to bribe the employees of its rivals in order to find out what they were doing. This could often be done through a very junior employee, and the cost was surprisingly small. In 1893 Standard's Philadelphia subsidiary, Atlantic Refining, paid a Negro clerk at the neighbouring Lewis Emery Oil Company a total of $90 spread over several months, in return for which it received full details of the quantities, buyers, and destinations of Lewis Emery's sales. When the man was eventually found out and sacked Atlantic did not even give him a job.†

The small companies could only really have challenged Rockefeller's position by building a pipeline system of their own, and this would have cost far more than they could afford. Moreover, whenever anybody tried, Standard would buy up tracts of land across the proposed route, and the rail companies would not allow the line to pass under their tracks. If the builder still persevered Standard had other methods to fall back on. When the 23-year-old William Larimer Mellon, of the Pittsburgh banking family, secured a court order allowing him to lay a line across Pennsylvania regardless of whose land was in the way there was a fight every time he had to go over the Pennsylvania Railroad's right of way. On several occasions when a stretch had been completed gangs of thugs would rip it up during the night.‡ Mellon, with his family's fortune behind him, persevered, and in 1892 he completed the project at a cost of over $1m. The effort was well

---

* Ida Tarbell, op. cit.   † Ibid.
‡ *Since Spindletop*, Craig Thompson. Gulf Oil Corp.

worthwhile. As usual when it was up against a dangerous adversary, Standard tried money where force had failed, and Mellon accepted an offer of $2½m. to sell his line.

The ruthlessness of Standard's competitive practices provide a striking contrast with the rest of Rockefeller's life. Whenever he appeared in court or before an investigating committee people were surprised by his benign and rather humdrum appearance. Yet it accurately reflected one side of his character. Despite his enormous wealth and his houses in different parts of the country, his way of life remained basically simple, revolving as it did around his family and the church. He never splashed his money around on ostentatious living like so many of those who made their fortunes in the same period. Instead he devoted his considerable energy and ability to building up one of the most efficient and free-spending philanthropies the world has ever seen.

He once told a friend: 'I had no ambition to make a fortune. Mere money-making has never been my goal. I had an ambition to build.' * And this is what he did. At the end of his life he was proud that he had brought order out of chaos in the oil industry, and created one of the strongest and most efficient business organizations to be found anywhere. He never seems to have seen a dichotomy between his private and business lives. To us this seems incredible, but nineteenth-century American business was a lawless jungle in which men fought with whatever weapons they could lay their hands on. Rockefeller's methods were certainly no better, but equally certainly no worse than those of such contemporary tycoons as Vanderbilt, Armour, or Morgan. We can condemn the morals of the period, but we must judge the individuals by the standards of their day, just as we would a Roman emperor or a Red Indian chief.

* Nevins, op. cit.

CHAPTER 4

The break-up of the Trust

The first hint of danger to Standard's domination of the world's oil came from Russia. At Baku in the Caucasus oil had been seeping through the earth's surface for hundreds of years and providing the fuel for the sacred and eternal fires of the Zoroastrians. Pilgrims came from all over Persia and India to worship at the shrine, but the commercial possibilities of the deposits were not grasped by the Tsarist régime until 1873, when it was decided to allow private prospectors into the area. Foreigners as well as Russians took advantage of the opportunity, and among the earliest pioneers were three Swedish brothers – Ludwig, Robert, and Alfred Nobel.

The Nobels have been described as 'the most inventive family of all time',* and though this claim must be difficult to substantiate, their combined achievements were undoubtedly remarkable. Their father Emmanuel invented torpedo boats; Ludwig and Robert built the St Petersburg dockyards, and Alfred invented dynamite. It was also he who founded the Nobel prizes for peace, literature, and scientific achievement.

The brothers quickly established an ascendancy over Baku, and, following Rockefeller's example, secured a virtual monopoly over the transportation of oil from the Caucasus to the rest of Russia and abroad. This was a complicated process, as there was no railway to the Black Sea, which was the logical outlet for exports from Baku. The oil had to be taken up the Volga and thence by rail to the Baltic. The other operators tried to build a railway to the Black Sea port of Batum, but the Nobels promptly slashed the price of their oil so low that it was impossible to raise the necessary finance. Only if money could be raised outside Russia could the

* *Mr Five Per Cent*, Ralph Hewins. Hutchinson.

project be undertaken, and accordingly in 1880 the Russians approached Baron Alphonse de Rothschild in Paris.

Although banking was the Rothschilds' main interest, the Paris branch of the family was deeply involved in oil. Baron Alphonse imported supplies from the United States, and owned a refinery at Fiume on the Adriatic which was well placed to handle a large volume of Russian imports. It was as bankers that the Rothschilds agreed to put up the money for the Batum railway, but in return they demanded mortgages on the producers' properties, and the right to buy oil for export.

With the completion of the railway in 1883 the growth of the Russian oil industry was incredible, and by 1888 its production was over 2½m. tons a year. The entire field at Baku covered only a few square miles, instead of being scattered over a wide area as in Pennsylvania, and the yield from its wells was infinitely greater than those of the United States. Some of the more spectacular 'spouters', as the most prolific wells were called, could throw their oil more than 300 feet into the air, and one of the wells belonging to the Nobels could produce more than a million gallons a day.

The extravagance of some of the rich producers would not have been out of place in the palaces of Kublai Khan. One built a palace out of gold plate, another modelled his on a house of playing cards, and a third constructed his storage tanks out of platinum. They imported vast quantities of beautiful women from all over Asiatic Russia, and maintained private armies of dispossessed Georgian nobles, called kotschis, to guard them from each other. The workers, meanwhile, were crowded into wooden barracks, and sometimes paid only in bread and water. The Nobles and other foreigners treated their employees far better than the Russians, but nowhere else, even in Russia, could the extreme of wealth and poverty have been more terrible, and it is not surprising that Georgia has provided so many Communist leaders, including Stalin and Beria.

None of this prevented exports from rising in line with production. Russian oil swept over Europe like a flood and transformed the entire marketing situation. In Germany, Galicia, and Rumania there had been a certain amount of crude oil production for some years, and in Britain, France, and Germany kerosene was also manufactured from shale, lignite, and coal. The Scottish shale companies were fighting back particularly strongly against the inroads of American oil, and in the early 1880s they were extracting as much as 30 gallons of crude oil from every ton of shale. None the less, nothing could have stood in the way of American oil, and Standard would have dominated Europe as thoroughly as the United States if it had not been for the arrival of Russian oil. In southern Europe, where a great oil port was established at Trieste, it took over the lion's share of the market, and even as far north as Britain it sometimes accounted for 30 per cent or more of the country's annual imports.

Europe became a kaleidoscope of cartels and price wars as Standard, the Nobels, the Rothschilds, various Russian independents, and the occasional American independent jostled for position. In general, they did not sell direct to the public, but through wholesale distributors and retailers. Simultaneously in some countries there would be all-out price cutting, while in others there would be a division of the market between a number of alliances, and in a third cartels including all the main competitors might be formed. The pattern was continually changing, and companies that were allied in one place could be at war in another. In theory, everybody was in favour of 'orderly marketing' and against competition on price. But a price war was often the only way in which an organization could increase its share of the market, and they were always fought with great bitterness as each contestant tried to secure contracts that would give it a strong bargaining position when the inevitable peace negotiations should begin.

By the 1890s competition was almost as intense in the Far East. Russian oil could be shipped there via the Suez Canal more easily than from America, and soon afterwards the Dutch East Indies became a major producer and Burma a less important one. By the end of the century there was nowhere left where Standard could be sure of earning high profits to offset any losses that might occur in the United States should new rivals appear on the scene.

And new rivals did appear. On 10 January 1901 the Texas oil industry was born when Spindletop, 'the most famous well in the world', blew in with an explosion that was heard for miles around and a jet of oil and mud hundreds of feet high. As John Bainbridge has said in his book about Texas:* 'That event, a turning point in our history, marked the end of the oil monopoly and the beginning of the liquid fuel age.' Before Spindletop Standard's position looked as secure as ever, but within a year of its discovery it was having to contend with a new breed of competitor, some of which are now among the most powerful companies in the world, such as Gulf and Texaco (officially called the Texas Company from 1902 until 1952).

Spindletop was a marshy hillock about a mile south of the town of Beaumont, and the story of its discovery is one of the saddest in the industry's history. The man primarily responsible was an individual with one arm and the colourful name of Patillo Higgins, who started life as a lumberjack and in due course established himself as a property dealer. He became convinced that Spindletop contained oil, but there were no surface indications, such as seepages or outcrops of oil-bearing rock, to support this view and all the experts told him that this was impossible. Nevertheless, Higgins was sure, and for ten years he devoted his life and $30,000 to trying to prove himself right. When he could raise no more money he persuaded Anthony Lucas, a former captain in the Austrian Navy who had emigrated to the United States, to take over the project. Lucas, too, spent all his resources in vain, and was forced to sell out to a syndicate backed by the Mellon brothers, the Pittsburgh bankers who a few years earlier had built a pipeline across Pennsylvania. It was this group which eventually located the oil.

Lucas managed to gain some credit from the discovery, and retained a small stake in the venture. Afterwards he made a decent living as a mining engineer. But poor Higgins was less fortunate. His pioneering work was completely forgotten, and for most of the rest of his life he roamed the state looking for another bonanza. He never found it, and when he died in 1955 at the age of ninety-two he was a poor and forgotten man. His only consolation was to receive a document signed by thirty-two leading citizens of Beaumont and attested by the County Clerk, which read: 'Mr Higgins

* *The Super Americans*, John Bainbridge. Doubleday.

deserves the whole honor of discovering and developing the Beaumont oil field. He located the exact spot where all the big gushers
are now found.'*

The gushers were unlike anything that had been seen before.
In a single year the first Spindletop well alone produced as much
oil as 37,000 wells in the Eastern states. A host of companies
sprang up to develop the new reserves, most of which were as dis-
organized and short of capital as those which had tried to challenge Rockefeller in the oil regions of Pennsylvania. But a few
were on a much sounder footing. Texas already had a number of
rich men with fortunes made out of beef and ranching, and there
was no shortage of funds from outside the state. By a fortunate
coincidence this was also the period when Andrew Carnegie
and his associates sold out their steel interests to J. P. Morgan,
and much of the money paid by Morgan went straight into
Texas.

Spindletop was too good to last: on 10 August 1902 it stopped
gushing, and within a year its output fell from 62,000 barrels a
day to 5,000 barrels a day. For a while it looked as if the new
companies would collapse. But within a few months they were
able to find new fields elsewhere in Texas, and shortly afterwards
oil was also found in Louisiana and Oklahoma, which in those
days was still classified as Indian Territory.

The discoveries in the south-west coincided with the most
important changes to take place in the pattern of demand for oil
products since the industry's inception in 1859. For forty years
kerosene had been by far the most important product, and as such
the foundation of Standard's prosperity. But in the early years
of the twentieth century the rapid spread of electricity made a new
and better form of light available than the old oil lamps, so that
the graph of kerosene sales first began to falter, and then to
decline.

At the same time a vast new market opened up for petrol, †
which had hitherto been regarded as little more than a waste
product suitable only for a rather inefficient type of engine or

---

* Bainbridge, op. cit.
    † In the United States petrol is known as gasoline, which is also the word
which oilmen themselves generally use.

stove. In 1902 there were a mere 23,000 motor vehicles on the American roads, and sales of petrol amounted to 5·7m. barrels; by 1912 the number of cars and trucks had risen to over a million, and the demand for petrol was 20·3m. barrels;* and in 1914–15 more petrol was sold than kerosene, and it became the most important single product.† There was also an increasing demand for fuel oil to replace coal in factory boilers, locomotives, and ships. Most of the technological developments in these new areas were made in Europe and Russia, but as so often happens it was the United States which took the lead in their commercial application.

The growth of these new markets gave the new companies an opportunity to outflank Standard's entrenched positions, and to expand far more quickly than if kerosene had remained the principal product. They also tended to be much quicker at grasping opportunities than the Standard executives, who had been brought up in a different age. By 1908 the Texas Company (now known as Texaco) had extended its operations from its home state into forty-three others, and Gulf, as the Mellons' organization was called, grew almost as fast.

Standard was in poor shape to face these multiple challenges. In 1897 Rockefeller handed over the administration of the company to John D. Archbold, and a new generation of executives was taking over the senior positions. As Rockefeller's protégés they were able and hard-working men, but instead of being able to concentrate on business matters, they were having to spend much of their time either in court or studying legal problems. Between 1904 and 1909 more than twenty state governments brought actions against Standard to have it ejected from their territories.

The age of economic *laisser-faire* was coming to an end, and public opinion was running strongly against the numerous large industrial trusts that had grown up since the Civil War to dominate American life. There were many reasons for this: the small entrepreneurs and businessmen were afraid for their future in a system based on Darwinian principles of the survival of the fittest,

---

* *Since Spindletop*, Craig Thompson. Gulf Oil Corp.
† *The Texaco Story*, Marquis James. Texaco.

and the agricultural communities of the South and West felt they were being exploited by the manufacturers and bankers of the North and East. These sentiments were naturally reflected in Congress and the state legislatures, and they were given added force by the fears of the politicians themselves. A few of the more far-sighted, such as Theodore Roosevelt, who became President in 1901, realized that economic power is the father of political power and that the American political system was in danger of falling under the domination of the great industrial barons in the same way as the kings of England had succumbed to their medieval barons.

The rising hostility to the trusts was fanned by revelations about how they operated. Books by the 'muck-raking' writers such as Upton Sinclair, Lincoln Stefens, John Spargo, and Ida Tarbell, and reports from the Government's Bureau of Corporations, for the first time exposed their activities to the public gaze. The list of scandals was endless. Americans learned to their surprise that conditions in their factories were often worse than on a slave plantation; that food and drugs were sold which were sometimes positively dangerous; and that respectable businessmen thought nothing of swindling on a vast scale. In a single year the Sugar Trust defrauded the Government of $4m. in customs dues, while the insurance companies made such enormous profits out of their premiums that they could pay their chief executives more than $100,000 a year.

By comparison with some of its contemporaries, Standard's offences were relatively minor. Its labour relations were exemplary, and the safety precautions in its refineries were far in advance of the period. Standard's competitive methods were certainly ruthless to the point of viciousness, but they were no worse than those of any other industry, and as a result of the new fields in the south-west its monopoly was in any case breaking down. None the less, it was the largest, oldest, and most well known of all the trusts, and thus in the most exposed position.

By the middle of 1907 the Government had seven actions pending against the company, the most important being in the Federal circuit court of Eastern Missouri, which demanded that the Standard group of companies should be dissolved. For three years the case poured profits into the pockets of the legal profession, until in November 1909 the court decided that Standard constituted an illegal combination under the terms of the Sherman Act, a measure passed in 1889 to 'declare unlawful, trusts and combinations in restraint of trade, and production'. The company appealed, but again it lost, and on 15 May 1911 the Supreme Court ruled that within six months all the major subsidiaries should be severed from the holding company and established as fully independent concerns.

Rockefeller resigned his nominal presidency of the Jersey holding company, and watched his creation disintegrate. By the end of the year it had been divided on a geographical basis into over thirty companies, each with a separate management. Their stock was distributed to the holders of shares in the former holding company on a *pro rata* basis, so that a common pattern of ownership continued and the companies retained their old names. As a result, competition between them was for several years a good deal more restrained than the court would have liked. But it was only a matter of time before the historic and personal bonds were broken. The dissolution was an untidy process, leaving the new companies as unbalanced concerns, some having too much refining capacity for their needs, and others markets they could not supply. By 1920 they had started moving into each other's territories, and by the 1930s they were competing against each other as vigorously as against everybody else.

On the whole, they have done very well. Standard Oil (New Jersey), usually referred to as Esso, is still the largest oil company in the world. Mobil * (the former Standard of New York) is in third position, and Standard of California, Standard of Indiana, and Continental are all in the top ten.

Rockefeller himself lived on to become one of the greatest philanthropists of all time. When he died in 1937 he had given

* This company has changed its name several times since 1911. Until 1931 its official name was Standard Oil Company of New York and it was generally called Socony. In 1931 it merged with the Vacuum Oil Company to become Socony Vacuum. In 1955 it adopted the name of Socony Mobil, and in 1966 this was changed to Mobil. Whenever the company occurs in the text I have used the right name for the date.

away $550m. Although he was still a very rich man with a personal fortune of over $26m., none of it was derived from the companies carved out of the Standard trust. His sole remaining holding in a Standard company was a single share in the California concern, which he had retained for sentimental reasons because it was Certificate No. 1.*

* Nevins, op. cit.

CHAPTER 5

Deterding's triumph

While the American Government was campaigning for the dissolution of Standard, another oil empire was being created in London and The Hague that could tackle even Rockefeller's on level terms. It was formed in 1907 by the amalgamation of two of the largest European companies, the Royal Dutch under Henri Deterding and Marcus Samuel's Shell, into the Royal Dutch Shell Group.

Their resources extended across the globe, and the new group was the most international in scope that the world had seen. Its oil was drawn from fields in Russia, Eastern Europe, and the Dutch East Indies (now Indonesia); its sales network covered Europe, Africa, Asia, and Australia; its tanker fleet was more numerous than the merchant marine of all but the largest countries. The Dutch were the senior partners with 60 per cent of the shares, and with the completion of the merger Deterding became the dominant figure in the international oil industry, a position he was to hold until shortly before his death in 1939.

With his flamboyant and swashbuckling character Henri Deterding was the complete antithesis of the grave and sober burghers portrayed by Rembrandt, who are supposed to personify the Dutch character. He might have been descended from one of those *bon viveurs* whose glorious mistresses were painted by Rubens. Certainly he shared that artist's enthusiasm for beautiful and voluptuous women. His *amours* were famous throughout the capitals of Europe, and he sometimes rewarded his favourites with an impetuosity and generosity that outran even his ample financial resources. On one notable occasion in Paris he gave Lydia Pavlovna, the wife of a White Russian general, a £300,000 parure of emeralds from Cartiers, only to find that he could not pay for them until the arrival of his director's fees a few months later. Fortunately the jewellers agreed to wait, and soon afterwards Lydia became his wife. The ostentatiousness of her new husband's life, with his showy cars, country houses, and special trains, must have given her many happy reminders of the pre-revolutionary Tsarist court.

But Deterding was no playboy. Till the end of his life he would take a swim in cold water before breakfast, often breaking the ice to do so, and he was once overheard to say: 'If I were a dictator I would shoot every idle man.'* Notwithstanding his own hedonistic pursuit of pleasure, he always managed to work incredibly hard, and despite its enormous size and rapid growth, he continued to take all the major decisions in Shell until the mid-1930s. He was the type of businessman who, apart from his animal pleasures, finds complete fulfilment in his work. 'I look upon oil far more as a recreation than as a business; I have been so long in it that it amuses me, it is a kind of sport, and for the rest I do not care very much,' he once told a Royal Commission.†

His attitude towards the way in which his business should be run owed much more to Holland than his character. There is an old Dutch proverb, which says: 'Cooperation gives power.' It derives from the experiences of the various Dutch provinces in the six-teenth century when they were fighting for their independence against Spain, the strongest country in Europe. They realized that if they remained apart from each other they would be defeat-ed one by one, and that their only hope of survival lay in present-ing a united front to the enemy. Deterding believed that the same was true in the international oil industry, where Standard took the place of Spain.

Even after Standard was broken up and Shell had become one of the most powerful companies in the world, Deterding still hankered after a system of limited competition. Unlike Rocke-feller, he had no desire to create a personal monopoly, or to drive all his rivals out of business. But he did want the industry to be run in an orderly, and above all a profitable, fashion. He believed that

* Ralph Hewins, op. cit.
† *Royal Commission on Fuel and Engines, 1912.*

this could only be achieved if the companies would agree to work together in various ways, and to abide by a set of rules restricting their freedom of action.

He was particularly scornful of price cutting: 'You may call it a refined form of throat cutting, a stranglehold, a dog fight, or by any more appropriate or opprobrious name you like.'* Deterding argued that quite apart from lowering profits, price cutting also reduced the demand for oil. He believed that when prices are stable the customers will always maintain ample stock, whereas if they think that prices may fall they will buy as little as they can, on a hand-to-mouth basis.

In 1907 he put this view to John D. Archbold, who had taken over from Rockefeller as the executive head of Standard, although Rockefeller remained as president. Archbold replied that Standard believed just the opposite, and thought that fluctuating prices helped sales. Since Standard's own success was derived from ruthless price cutting and ferocious competition, it is not surprising that Deterding failed to make a convert. But the conversation was to have profound consequences for the future. Also present was a young man called Walter C. Teagle, who was destined a few years later to take over the leadership of Standard Oil (New Jersey), the largest of the units to emerge from the destruction of Rockefeller's creation. He was much impressed with what Deterding had to say, and during the 1920s and early 1930s the two men were to try to implement the ideas that Archbold rejected.

When the meeting took place Deterding was forty-one, and already head of the newly formed Royal Dutch Shell combine. Yet he had been in the oil industry for only eleven years. As a boy, after leaving school at the age of sixteen, he had become a bank clerk in Amsterdam; finding the life too dull, he emigrated to the East Indies. There he worked as a clerk in the Netherlands Trading Company, an organization engaged in both general trade and banking, eventually becoming manager of the Penang branch.

He might have remained in the Far East for the rest of his life had he not met J. B. August Kessler, the managing director of the

* *An International Oilman*, by Henri Deterding as told to Stanley Naylor. Harper and Brothers.

Royal Dutch. Despite its imposing name,* this was just one of a number of small companies that had recently sprung up to work the oil deposits of Sumatra. Like the rest, it was always short of money, and Kessler complained to Deterding that as the bank would not accept oil as security for a loan, he could never sell his output as fast as he needed money. Deterding suggested that the only way round the problem was to become a trustee of the bank. When the idea succeeded Kessler offered Deterding a job, and in May 1896 Deterding became the company's sales manager. He quickly proved himself indispensable, and when Kessler died in December 1900 Deterding succeeded him.

By this time the Royal Dutch was clearly established as the leading Dutch oil company in the Far East, but its position still seemed precarious compared with that of its main rivals, Standard and Shell. The operating conditions in the jungles of the East Indies were desperately difficult, and it was a constant struggle to raise enough money from Dutch sources to sustain the capital investment needed to build up a sales network capable of competing with the British and Americans. The company also had trouble with its oilfields. On one terrible occasion on New Year's Day 1898, while a party was being held in the presence of the local potentate, the Sultan of Langkat, to celebrate the launching of a tanker named after him, the management received a message that their most important field was producing salt water instead of oil. Kessler managed to buy in enough kerosene from Russia to keep the business going; but this was a desperate measure, and if new reserves had not been found in 1899 Deterding would have had to look for another job.

Throughout this period Kessler felt that an alliance with Shell might provide the best long-term solution to his company's problems, and during the 1890s he held several talks with Marcus Samuel. Deterding always opposed the idea. He knew that a union with Shell would result in a much stronger company, but he also realized that the British would be the senior partners. He therefore felt that the Royal Dutch should try to maintain its freedom of action until it was strong enough to negotiate an alliance from strength.

* The full name was The Royal Dutch Company for the exploitation of petroleum wells in the Netherlands Indies.

To everybody else the idea of the Royal Dutch dominating Shell must have savoured of *folie de grandeur*. While Deterding was still a bank clerk in Amsterdam Marcus Samuel was establishing himself as a leading figure in Far Eastern trading circles, and since 1892 had held a powerful position in the international oil industry.

Like Deterding, Samuel was a self-made man, but he started rather higher up the ladder. His father, the elder Marcus Samuel, had started life as a small Jewish merchant in the East End of London buying bric-à-brac from homecoming sailors. He was especially interested in seashells, and when these became the most fashionable ornamental motif of the mid-Victorian era with a value as great as many semi-precious stones he was able to build up a small trading empire with connexions all over the Orient. The young Marcus and his younger brother Sam were sent on a tour of the area on leaving school, and in 1878 they set up a trading company of their own.

The most rapidly growing market was Japan, where the westernization programme of the Emperor Meiji was creating an insatiable demand for manufactured goods of every kind. While Marcus remained in London, Sam established an office in Yokahama, and within ten years their partnership had become one of the leading foreign concerns in Japan. They provided the country with machinery, and in return imported shells, china, lacquer, and carvings into Europe. They also shipped Japanese coal all over Asia, and gradually widened their interests to include various other Far Eastern commodities such as tea, rice, and jute.

They conducted their business through a network of agents in the main trading centres, and instead of owning their own ships, they chartered vessels through a firm of London brokers called Lane and Macandrew. One of the partners was Fred Lane, who was also the London agent for the Paris Rothschilds' oil interests. He wanted to find a way of selling their Russian kerosene in the Far East, and some time between 1885 and 1888 he suggested to Marcus Samuel that the brothers should add it to their range.

The Far Eastern markets were dominated by Standard, and any newcomer trying to break into the area would become an easy victim to the kind of selective price cutting by which Standard maintained its supremacy in the United States. Only a company capable of establishing itself simultaneously in every market could hope to survive. Lane argued that with their existing network the Samuels could do this, but Marcus Samuel was sceptical. Because of oil's inflammability, kerosene could only be taken through the Suez Canal when packed in tins that were in turn packed in wooden cases, and it was carried by sailing ships rather like the great tea clippers. Both the tinplate, which came from Wales, and the wood were very expensive, and Samuel wondered whether it would be possible to match Standard's prices. Nevertheless, by 1890 he was sufficiently interested to allow Lane to take him on a tour of the Russian oil industry.

In the Black Sea port of Batum from which Russian oil was exported to the West Samuel found the answer to his problem. The Russians were the first people to develop sea-going tankers able to carry oil in bulk. They were desperately unsafe, and frequently blew up, but as soon as he saw them at work Samuel realized that they could completely revolutionize the economics of the Far Eastern trade. If only tankers could be designed to satisfy the safety requirements of the Suez Canal Company they could take kerosene through the canal in bulk, and then, after being steam-cleaned, return to Europe carrying rice, tea, or some other cargo. Under those circumstances his agents could easily undercut the Americans.

The Samuels always liked to check their new ideas with each other, and when Marcus returned to London Sam was sent out to confirm his brother's estimates. When he, too, returned with a favourable report they set to work to plan their *coup*. The first priority was to design a suitable tanker; the task was given to Fortescue Flannery, one of Britain's leading contemporary marine engineers, and by August 1891 the canal authorities had given his plans their blessing. With this guarantee the Rothschilds agreed to sign a nine-year contract, orders were placed for a fleet of ships, and the Samuels' two young nephews, Mark and Joe Abrahams, were sent to the East to build the storage depots where the tankers could discharge; for this responsible task, on which depended the success of the entire venture, they were paid a mere £5 a week,* a modest salary even by the standards of the time.

* *Marcus Samuel, First Viscount Bearsted and Founder of Shell*, Robert Henriques. Barrie and Rockliff.

While the preparations were in progress rumours of what was afoot leaked out. Standard, the Welsh tinplate manufacturers, and the shipping lines raised an outcry in Parliament and the Press; efforts were made to persuade the British Government and the canal authorities to impose a veto; and for a while it looked as if the venture might be stillborn. But Samuel went on building his ships, and by 1892 the danger had passed; the canal needed the extra revenue, and the British Government saw no reason to intervene. In the following month the first of the tankers, the 4,000 tons *Murex*, sailed from the shipbuilder's yard in West Hartlepool for the Black Sea. On 24 August it passed through the canal fully loaded, and a few weeks later it discharged its cargo into Mark Abraham's bright new installations in Singapore and Bangkok.

The Samuels' cost calculations were proved correct, and the oil was put on sale at less than half the price of Standard's. The demand was expected to be insatiable, but despite its cheapness, nobody seemed to want it; there was one vital factor that had been left out of account. In Europe buyers were accustomed to bring their own containers in which to carry their kerosene away, and the Samuels expected the Asians to do the same. But in Asia the battered old tins in which Standard's oil travelled all the way from the United States played a vital role in the local way of life; they could be built up into houses, or refashioned into cups, plates, and buckets, or used for a thousand and one other day-to-day tasks. For many people the tins were more valuable than their contents, and kerosene could not be sold without them. So several more months had to be wasted while container factories were being built near the storage depot.

From the first it was Marcus Samuel who was largely responsible for the oil venture, while Sam concerned himself with other aspects of their business. It is still something of a mystery how he raised the necessary credit for the *coup*. By the end of 1893 eleven ships had been launched, and their total cost must have been considerable. However, he did not need any working capital. When he took delivery of the oil from the Rothschilds they presented a bill which did not have to be paid for three or four months; by that time the oil was sold and the money cabled back to Paris.

The ships were the only part of the enterprise which Samuel

actually owned himself. For the rest he operated a syndicate, known as the Tank Syndicate, in conjunction with his agents in the Far Eastern ports; all transactions concerning both the kerosene for the outward voyages and the cargoes for the return trips were conducted on behalf of a joint account of all the members. In this way Standard was prevented from picking them off one by one through selective price cutting or takeovers. It was not until 1897 that Samuel formed the Shell Transport and Trading Company to take over the entire operation, with the members of the syndicate becoming the first shareholders. The choice of name was a natural one; to commemorate the basis of his family fortune, Samuel had already given all his ships the names of different shells, and his oil was sold under the brand name of Shell.

After his spectacular entry into the international oil industry Samuel's sales expanded rapidly, and in 1895 he was strong enough to reject an offer from Standard, knowing that this would mean the Americans launching an all-out price war in an effort to ruin him. The attempt failed, and by the turn of the century Shell had become an extremely prosperous concern. At the annual general meeting on 21 June 1900 Samuel announced that in the previous year profits had increased by 60 per cent to £369,475, and the company's £100 shares were standing at over £300. Although kerosene in the Far East was still by far the main source of income, the company was beginning to build up a position in the new market that was developing for fuel oil in ships and on the railways; it had already landed an important contract with the powerful German Hamburg-Amerika line, and steps were being taken to establish a chain of bunkering installations at ports on the main trading routes.

When Samuel told the shareholders that he foresaw the day when the shares would reach £500 or even £1,000 there was a good deal of laughter, but the optimism seemed well founded. Yet when Deterding took over from Kessler as the executive head of the Royal Dutch a few months later he was not being fanciful when he looked forward to dominating the British company. For those who were able to look behind the impressive figures Shell was a good deal weaker than it seemed.

Its most important weakness was the character of Samuel

himself. Unlike Rockefeller or Deterding, he was not a dedicated businessman willing to devote his life to the growth of his company. His greatest ambition was to achieve status and acceptance in those circles from which his family had always been excluded. He wanted to take the Samuels out of the East End and to establish a dynasty of country gentlemen in the classical English tradition; as soon as his oil venture started to pay off he bought a large country house – The Mote near Maidstone – sent his son to Eton, and became a city alderman. From then on public service and civic affairs took up an increasing amount of his time, and he came to regard oil as being merely the means by which he could acquire the money to fulfil his social ambitions.

As a result, he did not have the energy to follow through his initial *coup* by building up a proper organization. Shell was run in an incredibly amateurish fashion; it did not even have a staff of its own and was run more or less as a department of the original M. Samuel & Co. trading concern along with the seashells and everything else. The effects of this weakness at the centre were reflected in Shell's world-wide strategy. Throughout the 1890s, despite its growing profits, it remained almost entirely dependent for supplies on the contract with the Rothschilds.

Samuel made only one half-hearted attempt to find reserves of his own in the Dutch East Indies. The casual fashion in which the operation was mounted was symptomatic of the company's whole approach. The untrained Mark Abrahams was sent out to Kutei in Borneo to take charge after a mere two weeks sightseeing in the Russian fields. When he none the less managed to find oil in 1897 all the equipment sent out from London was based on Russian specifications and took no account of the different characteristics of the Kutei oil; in the end Abrahams could only get his refinery to work by running it backwards.* By then it was in any case clear that the oil was unsuitable for refining into kerosene. Samuel was left as reliant as ever on the Rothschilds, and by 1899 it was clear that he would have to re-sign his contract.

To disguise his weakness he decided to put on a show of strength

---

* The Kutei oil contained an unusually high percentage of toluol. This is an essential ingredient in TNT and made the oil invaluable during the First World War.

by ordering as much as possible from other Russian companies, and promising to deliver large quantities to buyers all over the world. With the outbreak of the Boer War oil markets boomed and it looked as if the gamble might succeed. In October 1900 the Rothschild contract was duly re-signed, but then the bubble burst. The Boxer Rebellion and the Siege of Peking ruined the Chinese market, where Shell had built up large stocks, while in Russia the onset of a slump forced the refineries there to switch sales to the export market. As the surplus mounted prices everywhere collapsed.

Samuel's position might have been hopeless, had it not been for the discovery of the great Spindletop field in Texas in January 1901. As soon as he heard the news he opened negotiations with the producers, and prepared for another *coup* to make Shell as powerful in Europe as it was in the Far East. A massive contract involving a minimum delivery of 10,000 tons a year was signed with the Mellons' company, and Fortescue Flannery was called in to design four new 10,000-ton tankers to bring the oil across the Atlantic. In England Samuel hoped to persuade the Royal Navy to use Texas oil as fuel, while in Germany he purchased a large importer and formed a partnership with the Deutsche Bank.

Shell looked stronger than ever, and Standard renewed its offer of a take-over. In December 1901, after several months of haggling, it suggested a price of £8m. (about £40m. in today's money) on condition that it should receive full control. Although this was more than double the amount of money invested in Shell, Samuel refused; he did not want his creation to become just another cog in the great American machine.

Nevertheless, he was ready to hand over many of his responsibilities if only Shell could maintain its independence in alliance with a suitable partner. He realized that the oil industry was becoming too complicated to be run as an adjunct to a trading concern, and he wanted more time to concentrate on public life. The following November he was due to become Lord Mayor of London; no Jew had ever held the position before, and it meant the fulfilment of all his social ambitions.

Deterding saw his chance. He had watched the courtship between Shell and Standard with mounting concern; if it had led to

(a) Calouste Gulbenkian – 'Mr Five Per Cent'

3. (b) Marcus Samuel

3. (c) Henri Deterding

4. (*a*) Filling stations: old

4. (*b*) Filling stations: new

marriage the combination would have been strong enough to sweep everybody else out of the Far East, and the Royal Dutch's ambitions in Europe would have been stifled at birth. The only sure way of removing the danger was for the Royal Dutch itself to go into partnership with Shell, and although it was still smaller than the English company, Deterding felt that it had reached the position where it could hold its own. Instead of approaching Samuel direct, Deterding was shrewd enough to enlist the services of Fred Lane as a go-between. They began putting proposals to Samuel even before the negotiations with Standard were concluded, and the knowledge that the Royal Dutch was in the background must have influenced Samuel's rejection of the American offer.

It soon became apparent that as they did not intend to merge completely Shell and the Royal Dutch would have to establish a joint refining and marketing company to handle their respective supplies of oil. This gave rise to the key question of who should be the managing director, since the holder of that office would obviously play a crucial role in the affairs of both the parents. It was settled with surprisingly little difficulty early in April 1902 when Deterding and Samuel met for only the second time in their lives. Although the thirty-six-year-old Dutchman was thirteen years his junior and had only a fraction of his commercial experience, Samuel agreed that he should have the job. Samuel was already beginning to plan ahead for his Lord Mayor's inaugural procession in November, and he recognized that he was confronting a professional businessman 'whose purpose was to make money',* and who would devote all his considerable energies to the interests of the new venture.

He was, of course, completely correct in this judgement of Deterding's character, but he overlooked the fact that the interests of Shell and the Royal Dutch would by no means be always identical. In the first place their partnership was to apply only to Asia and Africa – in Europe, where Samuel was hoping to use his Texas oil to carve out a large share of the market, they would still be free to compete. Secondly, even within the agreement area, there would be many occasions on which the managing director

* *The History of the Royal Dutch*, F. C. Gerretson.

of the joint company would have to choose between the interests of its parents: whereas the Royal Dutch had plenty of oil, it was relatively weak on the distribution and shipping side, while Shell was in the opposite position. Under these circumstances the job should have been given to a neutral whom both sides could trust.

Not surprisingly, in view of Deterding's close connexion with Lane, it was decided that the Rothschilds should also be invited to join the new group in order to broaden its base. On 27 June the three parties agreed to set up a British company called the Asiatic Petroleum Co., with the power to impose production controls on the parent organizations, and to lease all their refineries and installations in the East. No binding undertakings were given to anybody. It was simply agreed that a company should be set up; it was not in fact established. Yet Deterding immediately started to behave as if it had been and to carry out the entire business under his own name. Samuel acquiesced in this astonishing situation and arranged a large personal overdraft for him at a British bank before leaving for the country to forget about oil while preparing a report on the state of the docks for the Port of London Committee. The Dutchman grasped his opportunity with both hands, set up an office in Billiter Street next door to Standard's local subsidiary (Anglo-American), and went to work.

Even while the negotiations leading up to the Asiatic agreement were in progress, the balance of power had started to slip away from Shell. On the day before agreeing to the admission of the Rothschilds Samuel went down to Portsmouth to watch H.M.S. *Hannibal* carry out the Royal Navy's trials on fuel oil. The fuel was supplied by Shell, and guaranteed to be smokeless, among its many other advantages. Samuel was confident that it would be conclusively shown that oil was in every way superior to coal and that Shell would become the official suppliers to the Navy. But the event was a disaster; the ship had been fitted with the wrong sort of vaporizer burners and as soon as it turned on its engine it was enveloped in dense clouds of smoke and soot. In the eyes of the Admiralty oil was damned, and Samuel's hopes of capturing the largest market in the world disappeared.

In August, while he was resting in the country, there came news of a still worse disaster when Spindletop stopped gushing. Within a few months Shell's supplies from Texas were completely cut off, leaving the company with large contracts to fulfil and no means of doing so. There was no longer any work for the 10,000-ton tankers, the pride of its fleet, and they had to be converted into cattle carriers before they could find employment.

Yet Samuel continued to concentrate on public affairs to the exclusion of his business responsibilities. In November 1902, when the Rothschilds and the Nobels sent top-level representatives to London to suggest a European marketing alliance to handle Galician oil, which could have filled the gap left by Spindletop, he was too busy to receive them. The emissaries were kept waiting six days while Samuel went through the Lord Mayor's ceaseless round of banquets, speeches, and dressing-up and the chance was lost. Shell continued to buy expensive Russian and Far Eastern supplies to meet its European commitments, and three years later in 1905 it paid the price of Samuel's casualness. It was forced into a humiliating withdrawal from the continental markets when it had to sell six of its best tankers to the Deutsche Bank as the price of getting out of their partnership agreement.

Meanwhile in Asia the Royal Dutch went from strength to strength as Deterding took over the smaller Dutch companies and formed the others into an association under his leadership. With ample supplies of cheap crude oil behind him he was able to move into Europe as Shell moved out, and in the internal transactions of the Asiatic he always favoured his own company. For a while Samuel considered going to arbitration, and an adjudicator was even appointed before he backed down. But despite these grievances, he agreed to give the Asiatic a corporate status rather than break off the partnership so that Shell could continue on its own.

Although he wanted to dominate Shell, Deterding had no desire to annihilate it. The partnership provided him with his only prospect of operating under the Union Jack and of gaining access to the financial resources of the City on equal terms with a British company. Holland was by now too small for his ambition. Only the British Empire at the plenitude of its power could provide him with a worthy base. Therefore he did not want to see

Shell destroyed; he just wanted it to be reduced to the point where it would seek full amalgamation with the Royal Dutch on his terms.

By the end of 1905 Shell was in a desperate position. Its only hope of survival lay in obtaining new reserves and markets outside the Asiatic to enable it to deal with the Royal Dutch on an equal footing. Samuel went to the British Government for help, asking for concessions in Burma, where there were small deposits being worked by the Burmah Oil Company,* and for another chance to gain a fuel oil contract from the Admiralty. The Government refused to help. Both, he was told, were reserved for wholly British concerns, and Shell, with its reserves and other interests in the Dutch East Indies, was not British enough.

The company's last remaining assets were those held by the Asiatic; so Samuel turned to Deterding with a suggestion that the two of them should amalgamate. He proposed a 50/50 deal 'no party to have any advantage over the other'.† Deterding disagreed; knowing the weaknesses of Shell better than anybody, he insisted on a 60/40 arrangement in favour of his own company. For a few months Samuel searched around for an alternative, but none was available, and in April 1906 he had to accept Deterding's terms. To recover from his disappointment he then departed for a four-months cruise in his new yacht, the *Lady Torfrida*, leaving Deterding a free hand to put the merger into effect. By the middle of 1907 all the legal formalities had been completed, and the Royal Dutch Shell group was officially established with dual headquarters in London and The Hague.

The Royal Dutch and Shell were reconstituted as holding companies dividing all the shares, profits, and expenses in their combined assets on a 60/40 basis. In the joint management committee the Dutch naturally received a built-in majority, but Deterding was unconcerned with matters of national pride. As London was the world's commercial capital, he established his own office there, and because he preferred the Shell brand name to that of the Royal

---

* Formed in 1871 and originally called the Rangoon Oil Co., Burmah was incorporated in 1886 in Scotland by David Sime Cargill, a Glasgow merchant with Far East business interests.

† Henriques, op. cit.

Dutch it was adopted for all the group's products. As a result, the group soon came to be thought of as an entirely British concern.

The amalgamation was carried through at that crucial moment in the oil industry's history when the general public was just beginning to buy cars in large quantities. In 1909 the demand for petrol in Britain alone was over 40m. gallons a year, and by 1912 there were more than a million cars on the roads of the United States. Not since the early 1860s had the oil industry faced such a promising future with so many opportunities for expansion.

Deterding set out to grasp them all. To meet the rising European demand he developed new fields in Rumania, and in 1913, when the Rothschilds decided to dispose of their Russian interests, he bought these too. Nor was he content to confine his activities to the eastern hemisphere; wherever he went, he found himself up against competition from American concerns notwithstanding the fact that at home they had the largest market in the world. He determined to challenge them on their own ground; in 1912 he started selling in the United States, and two years later he began buying oilfields there.

The significance of this did not escape the American companies, but they failed to realize that Shell in its new form was an entirely new phenomenon in the international oil industry. Standard had derived its strength from the enormous reserves in the United States which could be shipped to any part of the world. Rockefeller never felt the need to find other sources overseas, and could always deal with a smaller competitor by flooding his market and cutting prices. Shell, by contrast, had no large home base, and it was Deterding's policy always to secure a source of oil as near as possible to his markets. In this way he was able to cut down transport costs and to build up a chain of fields all over the world, which in the long run was bound to give him far more flexibility than his American rivals.

With the newly acquired interests in the United States it was natural that Deterding should also turn his attention to Mexico, where oil had been found soon after Texas and Oklahoma. Both the British civil engineer Sir Weetman Pearson (later the 1st Lord Cowdray) and the American prospector Edward L. Doheny owned such prolific reserves that their fields were called the Golden Lane; in January 1914 Shell, too, found a gusher. But in the long run it was Deterding's purchase for £1m. of the exploration rights in Venezuela belonging to a small American company called General Asphalt that was destined to become his most important pre-war exploration decision. Venezuela is now the largest single oil-exporting country in the world; when Russia expropriated Shell and the other Western companies after the Communist Revolution the Venezuelan fields saved Europe from becoming completely dependent on the United States, and more recently they have acted as an invaluable counter-weight to the Middle East.

CHAPTER 6

Gulbenkian's dream

With the benefit of hindsight it seems surprising that Samuel did not investigate the possibilities of the Middle East when he was looking for ways of breaking out of his dependence on the Rothschilds and the Asiatic. As early as 1871 a party of German geologists had visited the vilayets of Baghdad and Mosul in Mesopotamia* and brought back enthusiastic reports of the seepages being worked there by the local potentate, or Wali, Midhat Pasha. In Persia the prospects looked equally promising, and in 1872 Baron Julius de Reuter, the founder of the news agency, took out a mineral concession there.

Both areas were, however, wild and difficult, and the nature of their governments made it impossible to attract the large-scale investment that would be needed to mount an oil-exploration venture. The vilayets were under the rule of the Turkish sultan at Constantinople, while Persia was an independent empire with its own shah in Tehran. The Turkish Government was corrupt, arbitrary, and extortionate, and the shah was so weak that his writ was barely recognized outside his capital city.

It was left to the Germans to lead the way. One of the principal aims of German foreign policy in the late nineteenth century was the construction of a railway from Berlin to Baghdad, and in 1888 the Ottoman Railway Co., controlled by the Deutsche Bank, secured concessions in Anatolia and Mesopotamia for the purpose. The terms included all the mineral rights for 20 kilometres along both sides of the projected track, and the bank soon became almost as keen on these as on the railway itself. There were plenty of seepages and other surface signs suggesting the presence of large subterranean oil deposits, and this evidence was backed up by

* Now part of Iraq.

historical records. The town of Hit on the Euphrates had been a
source of bitumen for 5,000 years, and Herodotus wrote about it
being used in the walls of Babylon; on another occasion Noah is
supposed to have caulked his ark with pitch, which could have
been crude oil, and Genesis mentions slime being used as mortar in
the Tower of Babel.*

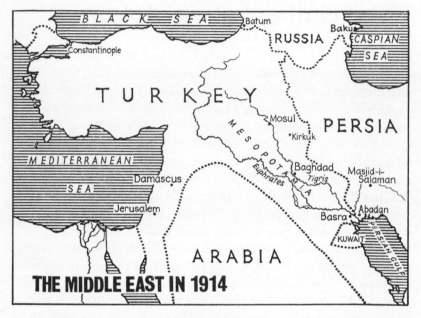

FIG. 3

The interest displayed by the Germans in oil aroused the suspicions of the Imperial Government. Hagop Pasha, the Minister
of the Liste Civile and guardian of the Sultan's private finances,
and Selim Effendi, the Minister of Mines, decided to find out
what all the excitement was about. If there was anything of value
to be found in the empire they felt that it should belong to the
Sultan rather than to a foreign concessionaire. There was nobody
in the Government with enough technical knowledge to offer a
clear-cut opinion either way, and no foreigner could be trusted to

* In *The Marsh Arabs* (Longmans, 1964) the explorer Wilfred Thesiger
describes how pitch from Hit is still used on canoes by the people who live in
the marshes of Southern Iraq.

tell the truth. They therefore sought the advice of a young man of twenty-one called Calouste Sarkis Gulbenkian, whose father was a prominent Armenian businessman in Constantinople.

Gulbenkian was something of a child prodigy; despite his youth he already had considerable experience of the world outside Turkey and an established reputation as an international oil expert. When only sixteen-and-a-half he had been admitted to King's College, London, as an undergraduate, and at nineteen he had graduated with a first-class degree in civil engineering, after writing a thesis about mining engineering with particular reference to oil. Before settling down again in Constantinople he went on a tour of the Russian oil industry, and wrote about his experiences in a series of articles for the *Revue des Deux Mondes*, a leading Paris intellectual magazine.

He treated the ministers' request with cavalier self-confidence. As he explains in his Memoirs: 'Hagop Pasha and Selim Effendi asked me to gather all the information I could get on the oil prospects of Mesopotamia . . . I elaborated a comprehensive report which was nothing else than a compilation of various travellers' books, principally of reports made by Colonel Chesney on the East India missions, and particularly from what I had heard from different engineers of the Anatolia Railways who had been in Mesopotamia.' Nevertheless, Hagop Pasha was impressed, and the governors of the vilayets were told to issue official certificates, known as tapous, transferring the lands to the Liste Civile without payment. The Germans found their rights over-ruled and could do no more than secure an assurance that they would be given priority over other foreigners.

Characteristically the Sultan, the famous Abdul Hamid (the damned), did nothing to exploit his new-found riches; the effort of expropriation exhausted the energies of his decaying administration. If Hagop Pasha had displayed the same vigour in developing the vilayets as he had in taking them over the history of the Middle East in the twentieth century might have turned out very differently from the way it has. The Turkish Empire would have been sufficiently strong before the First World War not to need an alliance with Germany. If it had then maintained its neutrality after 1914 there would have been no British support for an Arab uprising, and if one had still occurred the Turks would probably have been rich and powerful enough to put it down.

Gulbenkian received nothing for his pains. 'My boy, you ought to be very proud because you served the Treasury of His Majesty, and to serve His Majesty's Treasury is to serve your conscience,'* Hagop Pasha told him, and then sent him on his way. Gulbenkian went, but he took with him his knowledge, and the dream of one day profiting from the exploitation of Mesopotamian oil. He was destined to work towards fulfilling it in London and Paris rather than in Constantinople.

In 1896 Abdul Hamid launched the last and most ferocious of the Armenian massacres that disfigured Turkish history during the nineteenth century, and Gulbenkian was forced to flee the country. His journey into exile proved a turning-point in his career. On the boat which took him and his family away from Constantinople to Egypt he was lucky enough to meet Alexander Mantashoff, an Armenian Russian and one of the leading figures in the Russian oil industry. Gulbenkian immediately set about making himself indispensable to the great man. Although his wife Nevarte was nursing their young son, she was forced to give up her cabin for Mantashoff, and when they arrived in Egypt Gulbenkian stayed with Mantashoff as a sort of secretary-cum-butler.†

For a while the Gulbenkian family settled in Cairo, where Nevarte's cousin Nubar Pasha, for many years the Khedive's chief minister, was living in retirement. As his biographer, Ralph Hewins, explains: 'Gulbenkian was not one to ignore such an influential connection. He stuck as closely to the Pasha as he did to Mantashoff.' The Pasha was obviously impressed with his relation's ability and agreed to become God-father to the Gulbenkian's young son, who was christened Nubar in his honour. Gulbenkian himself became a sort of protégé which enabled him to build up a position of his own in society. This brought him into contact with the British Consul General, Sir Evelyn Baring (later Lord Cromer), with whom Nubar Pasha was very friendly, and through Sir Evelyn he was able to gain an introduction to the City of London, where Baring Brothers were one of the leading merchant bankers.

* *Memoirs.*   † Ralph Hewins, op. cit.

Despite his influential connexions in Cairo, Egypt was too much of a backwater for Gulbenkian to stay there long, and in 1897 he moved to London. Through Mantashoff he was able to secure a position as the representative of a number of Russian oil producers, and within a few years was a prominent figure in London oil circles. Once again he knew who to choose as friends, and among his closest associates were Fred Lane and Henri Deterding.

It was at this point that Gulbenkian made one of his few mistakes. In 1900 he was approached by the Director General of the Persian Customs, General Antoine Kitabgi (another Armenian incidentally), with a suggestion that he should take out an oil concession in Persia. He sought Lane's advice, and when Lane recommended him to turn it down he did so. Within a few years vast reserves had been discovered in the area in question, but by that time it was in the hands of a man called William Knox D'Arcy, the founder of British Petroleum.

In 1900 D'Arcy had recently returned to England after making a fortune out of gold-mining in Australia, and seemed bent on dazzling London with his new-found wealth. He had the only private box at Epsom, apart from the royal family's, and at his town house in Grosvenor Square he entertained on a grand scale; on one memorable occasion Caruso and Melba both sang to his guests on the same evening, and he never tipped servants less than a gold sovereign. With his walrus moustache and enormous girth he looked the typical hedonist of the Edwardian era, but like a good Edwardian he believed that money should be put to work as well as spent. When he heard Kitabgi's proposition he was suitably impressed, and within weeks his secretary, Alfred Marriott, and a geologist had been dispatched to Persia to clinch the deal.

When Marriott arrived in Tehran in 1901 Persia was at the nadir of its fortunes. The British minister, Sir Arthur Hardinge, compared it with a 'long mismanaged estate, ready to be knocked down at once to whatever foreign power bid the highest or threatened most loudly its degenerate and defenceless rulers'.* The northern provinces came under a Russian and the southern a British sphere of influence, and over wide areas local tribes plundered almost at will. The shah, Muzaffar ed-Din, was in debt

* *A Diplomatist in the Middle East*, Arthur H. Hardinge (London).

to both these great powers, and as the privy purse and the national treasury were one and the same thing, these were personal debts.

The imperial ministers dealt with Marriott more as if they were the servants of an impoverished and spendthrift landowner wishing to raise money on his estate than as the representatives of an independent state. They had little idea of oil's value, and no interest in its future. They were concerned more with getting money out of D'Arcy than in imposing conditions on his operations, and they were successful in achieving their objective. Apart from bribes, the Government received £20,000 in cash, and the promise of a further £20,000 in shares plus 16 per cent of any future profits. In return, D'Arcy was given exclusive rights over 480,000 square miles of territory – the whole country apart from the five northern provinces near Russia – and complete exemption from all taxes. The Persians' 16 per cent share of the profits was rather generous compared with the 12½ per cent royalty usually paid to landowners in the United States, but both sides thought they had done well: the company because the arrangement would give them protection in bad years, and the Persians because they thought profits would always go on rising.

Drilling operations began in 1902 under the charge of G. B. Reynolds, a former official in the Indian Public Works Department, who had once drilled for oil in Sumatra. The conditions were even worse than those he had had to contend with in the Far East. During the summer the temperature reached 110 degrees Fahrenheit in the shade before seven o'clock in the morning; all the equipment had to be sent from England via Basrah and Baghdad, which meant infinite delays; and the local chiefs, who barely recognized the Shah's suzerainty, had to be constantly bribed with protection money. Eventually, in January 1904, after a plague of locusts and an outbreak of smallpox in the neighbouring settlement, where the expedition got its food and water, oil was struck. But the excitement was short-lived. In May the well dried up, and the venture was back again to square one.

By this time D'Arcy's position was becoming embarrassing. He had already lost £225,000 of his own money, his bankers were growing restive, and all efforts to raise further funds had failed. Doubts about putting money into the Middle East were as strong as ever, and the financiers 'would have nothing to do with the matter till oil is found'.*

It was the British Government that came to the rescue. Despite the fiasco when oil fuel was tested in H.M.S. *Hannibal*, there was a growing number of officers in the Royal Navy who believed that eventually oil would have to take over from coal because of its technical superiority. They were led by the redoubtable Admiral Lord Fisher, and as soon as he became First Sea Lord in 1904 an Oil Committee was appointed under the chairmanship of the Civil Lord of the Admiralty, Mr E. G. Pretyman, to look into the matter. In the view of Fisher and Pretyman it was vital that a source of oil should be discovered by a British company within a reasonable distance from Europe in an area that Britain could dominate. Persia was not part of the empire, but it looked ideal, and one of Pretyman's first acts was to introduce D'Arcy to the Burmah Oil Co., which though small had acquired a good deal of experience in Burma. In May 1905, with Pretyman acting as marriage broker, the two sides agreed to form a joint enterprise, with Burmah promising to provide the necessary working capital and to develop any field that might be found.

In Persia, too, the British Government tried to help, but with somewhat less happy results. When the expedition's camp was moved to a new location, and the local inhabitants, the Bakhtiaris, proved even more recalcitrant than those at the previous site, the consul suggested calling for a gunboat. The Admiralty responded by sending H.M.S. *Comet* up river to the scene of the disturbance, but before it could be of any assistance it ran aground on a mud bank.

The exploration was equally unsuccessful. Two more dry wells were completed, and in January 1908, when drilling began at a place called Masjid-i-Salaman (the Mosque of Solomon) in the Zagros Mountains, it was obvious that the expedition's days were numbered, unless something could be found quickly. In May, with the well down to 1,100 feet without success, the blow fell when

---

* *Adventure in Oil, the Story of British Petroleum*, Henry Longhurst. Sidgwick and Jackson.

Reynolds received a cable telling him to return home,* as the company's funds were exhausted.

It was too much to bear. After six years of exploration he felt that he must at least continue until the confirmatory letter arrived from London, which meant another four weeks grace. Only two were necessary. At 4.30 a.m. on 26 May 1908 a jet of oil fifty feet high shot out of the top of the derrick, and smothered the drillers† – a commercial field had at last been found, and the oil industry of the Middle East was born.

In England the excitement was terrific, and a new company called Anglo-Persian‡ was formed to exploit the discovery, with Burmah putting up £1m. for the ordinary capital and the public taking the Preference shares. In Glasgow, where the shares were issued, the enthusiasm was so great that a queue of applicants five and ten deep struggled for places at the bank counter, and among the most pressing of the distinguished figures asking for special treatment was Field Marshal Lord Kitchener, who had been told about the oilfields by a *Daily Telegraph* correspondent just back from Persia. D'Arcy was reimbursed for all the money he had spent and given shares in Burmah valued at £900,000 as a reward.

Once more, however, Persia was to prove far more expensive than anybody had bargained for. A pipeline had to be laid to the coast 130 miles away and a refinery built before the oil could be exported. By 1913 not a drop had been sold and the company was facing another financial crisis.

It coincided with a critical debate over naval planning and strategy at the Admiralty, where Winston Churchill had become First Lord of the Admiralty in 1911. He wanted to build a new Fast Squadron of Battleships, which would not only be faster than anything the German Navy could provide but also more heavily armed. If coal was retained as the fuel it would be possible to

---

* In the official company history Longhurst quotes an extract from the diary of Lt. Wilson, who was in command of the exploration party's bodyguard, which in turn quotes from the cable. But no copy of the cable has been traced, and the mailed instructions were to carry on until reaching 1,600 feet before giving up.

† Longhurst, op. cit.

‡ In 1935 the name was changed to Anglo-Iranian and in 1954 to British Petroleum.

achieve the necessary speed, but only at the cost of sacrificing a turret to accommodate the extra boilers. Although the Royal Navy already had several oil-burning destroyers and submarines, there was still a large body of opinion which argued that capital ships should not be changed over on the grounds that in the long run it would be safer to rely on British coal than foreign oil, the supply of which might be cut off in time of war.

Churchill was convinced that the Navy had to change over to oil-fired engines for reasons of technical efficiency: 'The advantages conferred by liquid fuel were inestimable. First speed. In equal ships oil gave a large excess of speed over coal. It enabled that speed to be attained with far greater rapidity. It gave 40 per cent greater radius of action for the same weight of coal. It enabled a fleet to re-fuel at sea with great facility. An oil burning fleet can, if need be, and in calm weather, keep its station at sea, nourishing itself from tankers without having to send a quarter of its strength continually into harbour to coal, wasting fuel on the homeward and outward journey. The ordeal of coaling ships exhausted the whole ship's company. In wartime it robbed them of their brief period of rest; it subjected everyone to extreme discomfort. With oil a few pipes were connected with a tanker or the shore and the ship sucked in its fuel with hardly a man having to lift a finger. . . . The use of oil made it possible in every type of vessel to have more gun power and more speed for less size and cost. It alone made it possible to realize the high speeds in certain types which were vital to their tactical purpose.'*

None the less, Churchill disliked having to rely on a fuel that could not be produced in the British Isles, and he also disliked the fact that the Navy would have to buy most of its supplies from Standard and Shell. He felt that as both were foreign-owned they would put their own commercial advantage above British strategic interests, and he suspected them of overcharging.

He believed that if the Anglo-Persian could be brought under government control it could provide the solution to all these problems. A commission under John (later Lord) Cadman, professor of mining at Birmingham University and petroleum adviser to the Colonial Office, was sent out to Persia to investigate the

* *The World Crisis 1911-14*, Winston Churchill. Thornton Butterworth.

situation. When its report was encouraging Churchill was able to persuade the rest of the Cabinet to agree to the Government becoming a part owner of the company. The necessary Bill was introduced into the House of Commons in May 1914, and received the Royal Assent just six days before the outbreak of war.

The agreement is still the only one of its kind; the government took up £2m. worth of shares in return for a voting interest of just over 50 per cent* plus the power to appoint two directors with a veto on all strategic issues. Although this veto has never been used, and the company always places great stress on its commercial independence, there is little evidence to suggest that it has derived much advantage from its financial link with the Government, but as the largest all-British oil company it has undoubtedly received help. The advantages of this position were revealed even before the relationship was formalized, when the British Government became interested in the opportunities for exploiting oil in the Turkish Empire.

The situation in Turkey had been dramatically changed by the overthrow of Abdul Hamid in the Young Turk revolution. The Young Turks regarded themselves as being liberal in outlook, and accordingly looked to Britain for support. To take advantage of this opportunity, three London financiers, Sir Ernest Cassell, Lord Revelstoke, and Sir Alexander Henderson, established the National Bank of Turkey in 1910. Also on the board was Calouste Gulbenkian.

Despite his unpleasant experiences during the Armenian massacre of 1896, Gulbenkian had retained close links with Constantinople, and managed to become financial adviser to the Turkish embassies in London and Paris. With his Turkish connexions, his friendship with Deterding, and his own position in the Russian oil-export trade Gulbenkian felt that at last he was in a position to do something about Mesopotamian oil. He explains in his *Memoirs* that it took some time for his colleagues to realize the poor prospects for pure banking in Turkey. When this had at last

---

* When BP purchased the chemical interests of the Distillers Company in 1967 part of the payment was made in shares. As a result, the government's shareholding was reduced to slightly under 50 per cent. However, the government still has its two directors.

sunk in, 'I referred to a scheme for petroleum concessions and monopoly in Turkey.'

The bank rose to the bait, but before he could get down to the serious work arranging for the mounting of an expedition Gulbenkian had to seek a *rapprochement* with the Deutsche Bank. As long ago as 1890 the Deutsche Bank had become interested in the possibility of finding oil along the route of the proposed Berlin-to-Baghdad railway, only to have its concessions expropriated by Abdul Hamid after Gulbenkian's report to Hagop Pasha. At that time the Germans had been promised that they would take priority over all other foreigners, and they had no intention of surrendering those rights.

The Anglo-Persian also felt that it had a right to be considered. As well as spending vast sums in Persia, D'Arcy had tried to secure a concession in Mesopotamia as well, and the negotiations were handed over to Anglo-Persian on its formation. The company claimed that on the eve of the Young Turk revolution it had been about to sign an agreement, and that consequently its rights should take precedence over all other claims.

Gulbenkian decided to ignore the Anglo-Persian and to concentrate on bringing the National Bank and the Deutsche Bank together in an alliance that would also include his old friend Henri Deterding and Shell. The appropriate company was in due course set up in London in 1911, and in 1912 it took the name of the Turkish Petroleum Company (usually abbreviated to TPC). Half the shares were given to the National Bank, including 15 per cent held on behalf of Gulbenkian personally, and the remainder were divided equally between Shell and the Deutsche Bank. Not surprisingly, Anglo-Persian continued to press its claims, and the Turkish Government followed its traditional policy of playing one side off against the other.

The British and German Governments decided that a general settlement of all outstanding claims would have to take place before a concession could be obtained. Accordingly, a conference was called at the Foreign Office in London during March 1914. At the British Government's instigation the National Bank was forced to drop out in favour of Anglo-Persian, and Gulbenkian lost much of his bargaining power. When the shares were redistributed Shell and the Deutsche Bank held on to their 25 per cent, while the remaining 50 per cent were given to Anglo-Persian, though, because of his past efforts and his influence in Turkey, Shell and the Deutsche Bank agreed that each of them would give a 2½ per cent stake to Gulbenkian. It was in this way that he secured his famous nickname of Mr Five Per Cent.

All the parties to the agreement undertook not to have any oil interests in the Turkish Empire independently of the Turkish Petroleum Co., and on 28 June the Grand Vizir promised to grant it a concession covering the vilayets of Baghdad and Mosul. The way was at last open for a great Anglo-German enterprise in the Middle East. But before anything could be done the First World War broke out, and the Deutsche Bank's interest was taken over by the British Custodian of Enemy Property. Once again Gulbenkian's dream had to be put back into cold storage.

CHAPTER 7

War and peace

'Truly posterity will say that the Allies floated to
victory on a wave of oil.'

Lord Curzon

The war transformed oil from being a source of revenue for
tycoons and speculators into a vital industrial and strategic raw
material. Its importance was demonstrated in the first few weeks,
when the Paris taxis rushed troops to the Battle of the Marne,
and thereafter the allied armies became increasingly dependent
on the internal-combustion engine. By the end of the war the
British Army alone had 79,000 cars and trucks and another
34,000 motor bicycles in service compared with the 827 cars and
15 motor bicycles with which its expeditionary force entered the
fray; tanks took over from the cavalry as the spearhead of the
advance, while in the air the 'aviators' added a touch of glamour
to the squalor of the trenches.

Apart from Russia, with its Caucasian fields, Britain was better
placed than any of the other combatant nations. Despite being
more than half Dutch,* Shell co-operated fully with the war
effort, and gave the country access to its fields in the East Indies
and Venezuela. Britain could also draw on Lord Cowdray's
reserves in Mexico, Burma, and finally Persia, where Anglo-
Persian's refinery at Abadan came onstream just before the
fighting began. As if all this was not enough, the Royal Navy's
supremacy at sea meant that imports from the United States
could go on without interruption.

Germany and Austria-Hungary were much worse off. For a
while they were able to go on buying from the Americans, who
made large profits out of selling to both sides, but the British

* Holland was neutral.

blockade of enemy ports soon made this impossible. After that the Central Powers had to rely on their own small indigenous production and such oil as they could salvage from Rumania, where the wells and other installations were blown up in the face of their invading armies.

As the fighting wore on their lack of oil became an ever more serious handicap. Although fear was the main reason why Germany's oil-fired Navy was immobilized in port after the abortive fray at Jutland, a shortage of fuel added to its strategic disadvantages, while in the field her armies had to rely on horses to do the work of trucks and motor bikes; as these were killed off their communications and reinforcements were badly slowed up. Behind the lines the situation was much worse. In the factories production was reduced by the shortage of lubricants, people who still relied on kerosene lamps were left without light, and civilian transport was eventually brought almost to a standstill. By contrast, London's buses were never taken off the roads, and in the United States, even after it entered the war, private motoring went on without interruption.

The main oil-producing countries in 1913

	Tons
United States	33·0m
Russia	8·6m
Mexico	3·8m
Rumania	1·9m
Dutch East Indies	1·6m
Burma and India	1·1m
Poland	1·1m

France's situation would have been just as bad but for Britain. As far back as the 1870s the French had insisted on building their own refineries so that oil from the United States could be imported in the crude instead of as products. But they failed to develop their industry any further. Unlike their British and Dutch counterparts, French investors did not want to put money into risky exploration ventures abroad; they preferred 'safe' government and commercial bonds, especially Russian, and the country relied for its oil supplies on 'les trusts', as the British and American groups were called.

When the war showed how dangerous this could be, Deterding and Gulbenkian thought they saw an opportunity to become the architects of a new French oil policy. Neither of them had liked the Foreign Office Agreement on the Turkish Petroleum Company; Gulbenkian because his 15 per cent share had been reduced to 5, and Deterding because he was worried about the effects of the British Government's support for Anglo-Persian. Accordingly, Gulbenkian suggested to Senator Henri Berenger, the senator from Guadaloupe who was head of the Commissariat General aux Essences, that France should ask the British for the Deutsche Bank's quarter share in Turkish Petroleum. It was understood that Shell would look after it, and Deterding established a joint company with the Banque de l'Union Parisienne for the purpose.

This request cut right across the already complicated skein of British, French, Jewish, and Arab claims in the Middle East. But eventually a satisfactory agreement was reached at the San Remo conference in April 1920, when Britain and France divided the Arab territories between them as League of Nations mandates. This meant that the Arabs were nominally independent while in fact under the tutelage of the great powers. Mesopotamia, which shortly afterwards changed its name to Iraq, fell within the British sphere, and Syria went to the French. As part of the settlement the French were given 25 per cent of the Turkish Petroleum Co., and allowed it to build a pipeline through Syria to the Mediterranean.

Deterding's plans, however, failed to materialize. When Raymond Poincaré became Prime Minister he was adamant that France should have an oil company of its own, just as Churchill had been in Britain; Shell's venture was shouldered aside, and the job of safeguarding France's petroleum interests was given to Ernest Mercier, a leading industrialist. In 1923 he established the Syndicat National d'Etude des Pétroles, and in the following year it became the Compagnie Française des Pétroles. From the very first it had the promise of official support, and in 1929 the Government again followed Churchill's example by taking a direct equity stake, which in 1931 was set at 35 per cent, with a 40 per cent voting power.

The negotiations leading up to the San Remo oil agreement were kept secret from the Americans, and when they heard about the settlement a major row broke out between the Foreign Office and the State Department. Throughout 1919 British officials had been refusing to allow Standard of New York to resume its pre-war exploration activities in Palestine, while Shell geologists were at work in Mesopotamia. The department believed that Britain was planning to use its military presence in the Middle East to establish complete control over the region with the aim of excluding American interests from the promising oil areas, and San Remo confirmed this view.

In the argument that followed the State Department took the line that as the war had been won by all the allied and associated powers fighting together, they should have equal access to the fruits of victory. In reply, the Foreign Office claimed that although this view was all very well, it still gave the Americans no licence to trespass on rights which had been acquired before the war, such as those belonging to the Turkish Petroleum Co. The State Department retorted by saying that the Grand Vizir's promise to the company did not constitute a valid concession.

Beneath the diplomatic niceties both sides were driven by fear. The British, knowing that one of the most important props of their power and influence in the nineteenth century had been their indigenous coal and iron ore, were determined to get as much of the world's oil under their control as possible. With the collapse of the Russian industry in the 1917 Revolution 65 per cent of the world's production lay within the borders of the United States, and there was no reason to suppose that Britain would be able to draw on this source of supply in any future war. The Empire had to have its own reserves, and the Middle East was the obvious place to search for them.

To the Americans this looked like a sinister attempt to corner all the available supplies in order to hold them up to ransom. Despite their dominant share of the world's production, the Americans were convinced that their reserves were on the verge of running out. Between 1914 and 1920 United States domestic demand had risen from 210m. barrels a year to 455m., and it seemed impossible that the domestic fields could continue to sustain such a rate of growth.

In 1920 crude oil prices shot up by 50 per cent to three times their 1913 level, and Dr George Otis Smith, the Director of the United States Geological Survey, warned that the country would either have to use less oil or depend on foreign sources. To reduce consumption with almost 10m. cars and trucks on the roads was unthinkable, and foreign in this context meant British.

Already Shell was one of the largest oil companies in the United States. While the successor companies to the Standard Trust were sorting out their affairs and jockeying for position with Texas, Gulf, and the host of other new companies that sprang up to feed the growing demand for petrol, the Shell subsidiary had grown rapidly; by the early 1920s its distribution network covered most of the country and its fields were scattered over the continent from California to Oklahoma. In Mexico it was the biggest single concern after taking over Lord Cowdray's interests in 1919, and in Venezuela, which by sea is nearer the big east-coast ports than Texas, most of the more promising concessions looked to be under Shell's control.

If Britain and France were going to keep the United States out of the Middle East, Americans began to wonder how they would ever be able to withstand such a colossus. It looked as if the British were trying to do on a world scale what Rockefeller had done in the United States, and statements by British experts and publicists frequently bore out this view. One which received particularly wide publicity after being quoted in Congress was an article in the September 1919 issue of *Sperling's Journal*, a London publication, by Mr E. Mackay Edgar. 'The time is indeed well in sight,' he wrote, 'when the United States . . . will be nearing the end of some of its available stocks of raw materials on which her industrial supremacy has been largely built. . . . America is running through her stores of domestic oil and is obliged to look abroad for future reserves. . . . The British position is impregnable. All the known oilfields, all the likely or probable oilfields outside of the United States itself are in British hands, or under British management or control, or financed by British capital.'

To the congressmen this was a bloodcurdling prospect, and in May 1920 Senator Phelan of California put forward the almost 'socialist' proposal that a state-owned corporation on the lines of Anglo-Persian should be set up to look for oil abroad. The idea was rejected, but Herbert Hoover, the future President who became Secretary of Commerce in 1921, realized that the American industry had to present a united front to the world; in November of that year he persuaded seven of the largest companies – Standard Oil (New Jersey), Standard of New York, Gulf, Texas, Sinclair, Atlantic, and Mexican – to form a syndicate to represent American interests in the Middle East.

Whatever the future might hold, America at that time was still the most powerful single force in world oil, with plenty of bargaining counters. Among the most important was Anglo-American, the former Standard subsidiary; it maintained close links with the New Jersey Standard and was second only to Shell in the British market. It is not known whether the Jersey board ever actually threatened to cut off supplies, but they certainly considered the possibility, and news of that sort gets around.* In any case, the British and French, prodded by the ubiquitous Gulbenkian, were becoming increasingly anxious to get on with the development of Iraq, and a compromise was in the best interests of everybody. So when Sir Charles Greenway, the chairman of Anglo-Persian, asked the American syndicate in June 1922 to enter negotiations with a view to joining the Turkish Petroleum Co., his invitation was promptly accepted.

* Federal Trade Commission report on the International Petroleum Cartel, 1952.

CHAPTER 8

New discoveries

If the British and French had stood out against the Americans for another year the pressure on them might well have been relaxed. Throughout the oil industry's history men have been afraid of what will happen when their supplies are exhausted or they become dependent on somebody else. But the crisis never comes; whenever a new demand arises vast new reserves are found to meet it. The original Pennsylvania field was discovered when the demand for light and lubricants was outrunning the available supply and threatening the progress of the American industrial revolution; the Russian fields were developed when Europe needed an alternative to the United States; and the big discoveries in Texas and Oklahoma came just in time for the motor car.

The same thing happened again after the First World War. As the American fears of becoming dependent upon the British reached their height there was another series of major discoveries.

On 14 December 1922 Venezuelan Oil Concessions (VOC) was drilling in an abandoned well called Barrosa 2 near the shores of Lake Maracaibo. Nothing very exciting was expected to happen; the company just wanted to make sure that every possibility had been examined. Even when the bit struck oil it was not realized that a large field had been found. Then suddenly the flow accelerated, a jet 200 feet high flew into the air and the whole derrick was destroyed; five hundred men were brought in to stem the flood, but the well gushed on for nine days, during which nearly 1m. barrels went to waste. Although VOC was a British concern in effect controlled by Shell, several American companies, including Standard Oil of Indiana and Gulf, held adjacent concessions, and a host of others quickly came into the area. From 2m. barrels in 1922, all from Shell, Venezuela's production rose to 9m. in 1924, to 37m. in 1926, and to 106m. in 1928, when it replaced Russia as the world's second largest producer.

This rapid increase was not entirely due to the prolific nature of the wells. It also owed a great deal to the policies of the Venezuelan dictator General Juan Vincente Gomez, and to events that were taking place in Mexico.

In 1910 the Mexican dictator Porfirio Diaz had been overthrown, and from that time on the country was racked by disorder and civil war. There seemed to be only one point on which all factions were united; they hated the foreign companies Diaz had brought in to develop the country. Partly this was because the foreigners had supported Diaz to the last; more importantly, it was because they provided a focus for Mexico's feelings of frustration, and a target for its resentment against the United States. Oil companies are always loth to leave a country once they have made their enormous initial investments; they prefer to soldier on in the hope that better days will come. In Mexico, however, the situation changed dramatically in 1919 when salt water appeared in the wells along the Golden Lane. In those days this spelt disaster for a field. The companies continued to extract as much oil as they could from the unaffected areas, but the combination of geological and political uncertainty made them anxious to find an alternative.

They turned to Venezuela, where Gomez welcomed them with open arms. He liked to play the Americans off against Shell and other British interests when he thought he could do so with impunity, but whenever there was a danger of offending the industry as a whole he backed down. When the Development Minister, Gumersindo Torres, tried to enact a Petroleum Law which the companies thought too stringent he was sacked and replaced by a former company lawyer called Rafael Hidalgo Hernandez. Together with industry officials, Hernandez formulated the 1922 law, which gave the companies the most favourable political climate of any in the world at that time. Royalties were fixed at 10 per cent, except for fields under Lake Maracaibo or 100 miles from the coast, where the rate was $7\frac{1}{2}$ per cent. The companies were given various privileges, such as the right to occupy any lands they might need for roads, camps, and the like, and the right to import their materials free of customs duty. Against this they were obliged to provide hospitals and accident insurance for their workers.

The law with minor modifications remained in force until Gomez's death in 1935, since when it has been widely criticized as a sell-out to the companies, on the grounds that the comparable royalty in the United States was 12½ per cent, and that the large profits later made by the companies show that better terms could have been secured. It is, however, easy to be wise after the event. The critics tend to forget that when it was negotiated before the discovery of Barrosa 2 Venezuela was a geological unknown quantity, and had a notoriously unstable political history. Nobody had any idea how much oil it contained. The only certainty was that vast sums of money would have to be spent establishing facilities in the jungle and in the middle of Lake Maracaibo; only generous terms could have persuaded the companies to take such risks.

The criticism should be directed not at the law itself but at the way Gomez operated it. Even by contemporary Latin American standards he was notably corrupt, and the allocation of oil resources provided him with opportunities beyond the dreams of avarice. After his death his personal fortune was estimated at over $200m., most of which was derived from the oil companies, while many millions more went to his family and friends. To the companies this seemed a small price to pay for stability, a docile working force, and freedom from irksome government restrictions. So long as the money kept pouring into his coffers Gomez gave them *carte blanche* to run their concessions as they wished, even allowing them to build their refineries in the neighbouring Dutch West Indian islands of Aruba and Curacao instead of diversifying the Venezuelan economy. The oil industry became a virtual state within a state, and its executives forgot the lesson of Mexico that too close an identification with a dictator is bound to lead to trouble when he goes.

The companies did not, however, make the mistake of becoming dependent on Venezuela for all their supplies. Despite its proximity to the United States Atlantic-coast ports and the fact that its production costs were much lower than Texas, they continued to look for fields in North America. Notwithstanding all the gloomy forecasts about the continent running out of oil, they were more successful than ever before. In California, Wyoming, and many other places, as well as in the south-west, new reserves were uncovered. In fact, 1921 turned out to have been the last year of shortage; from 1922 onwards the United States produced more oil than its refineries could handle, and the refineries in turn pushed out more products than the market could absorb.

The culmination came in October 1930 when an old man called Columbus M. Joiner, usually known as Dad, was drilling on the Widow Daisy Bradford's farm in Rusk County, in East Texas. It was an area where the geologists of the large companies said that oil could never be found, and Dad's first two wells were dry. But he was able to raise money by selling shares to the neighbouring farmers and, with the help of Mrs Bradford and a local bank manager who worked on the rig in his free time, Dad persevered. As a young man he had made and lost two fortunes, and his success at finding new fields in unexpected places had earned him the title of King of the Wildcatters. Though over seventy, he wanted one more success so that he could retire in luxury. On 3 October the drill bit at last struck oil, and with a tremendous roar a vast black plume shot through the top of the derrick. Dad, his crew, and the shareholders danced and rolled in the spray, and the scenario was set for countless Hollywood movies about Texas.

But, unlike the movies, the story does not have a very happy ending. Dad created such a glut of oil that prices everywhere collapsed, and most of the production came from areas outside his control. His own lease covered only a few square miles, and even this proved difficult to sell, since his book-keeping was so chaotic that he could not prove his legal ownership. Not until Haroldson Lafayette Hunt, another notable wildcatter and one of the most famous poker players in the south-west, arrived in Rusk County could Dad find a buyer. In the circumstances Hunt's price of $30,000 in cash, $45,000 in short-term notes, and a guarantee of $1·2m. out of future profits* looked very generous.

* 'Just Plain H. L. Hunt', Tom Buckley, *Esquire*, January 1967.

In retrospect, however, it can be seen that Hunt secured one of the best bargains of all time. The field has already produced more than 3,500m. barrels and shows no signs of drying up. Hunt's profits are estimated to exceed $100m., and he is now reputed to be the richest man in America. Dad fell back on hard times, and retired to Dallas to live in a small house in a street with the appropriate name of Mockingbird Lane, where he died in 1947.

In addition to these new oil reserves the United States also learned how to make use of natural gas for the first time, thereby opening up yet another vast source of energy. Natural gas is part of the same hydrocarbon family as oil, and the two are usually found together, though they can occur separately. As far back as 1824 the town of Fredonia in upstate New York began using a local deposit for heating and cooking, but until the inter-war period it was generally regarded as valueless. Its economic potential was not released until the development of high-strength, thin-walled pipe and electric welding in the early 1930s made it possible to build long-distance pipelines to carry the large reserves of Texas and the south-west to the big industrial centres.*

* For an account of the development of natural gas see Chapter 22.

CHAPTER 9

The United States gains the advantage in the Middle East

It was against this changing background that the negotiations took place between the United States syndicate brought together by Herbert Hoover * and the Anglo-French interests over American entry into the Turkish Petroleum Company. When they began in 1922 it was thought that nothing less than the regulation of the world's last great oil reserve might be at stake; when they ended six years later the industry's main problem had already become one of trying to curtail excess production. The reality of a surplus had replaced the fears of a shortage.

The negotiations themselves stand out as a watershed in the industry's history. For the first time they brought together all the world's largest companies, and the French at least were quite clear what this implied: 'It was the beginning of a long-term plan for the world control and distribution of oil in the Near East.' †
The results were the famous Red Line Agreement, which set the pattern for the Middle East concessions of today, and the Iraq Petroleum Company, the first of the great joint ventures which form the basis of the major companies' control over most of the world's reserves.

Throughout the negotiations the British and French companies were supported by their governments, and at every turn the interests of Calouste Gulbenkian had to be taken into account. He was determined to hold on to the 5 per cent share in the

* See Chapter 7.
† From a Memorandum by the Compagnie Française, which is reproduced in the United States Federal Trade Commission's 1952 report on the International Petroleum Cartel.

project which he had been guaranteed by Shell and Anglo-Persian at the 1914 Foreign Office Conference, and with his experience going back to his 1890 report to Sultan Abdul Hamid he knew more about the whole question than anybody else. He had also become a major figure in the international oil industry in his own right. Nobody could quite say how; he never actually ran an organization or seemed to own any physical assets; yet in concession deals all over the world he had become an indispensable middle man. His grasp of the industry's politics, his knowledge of each company's strength and weaknesses, and his memory for the actions and promises of companies and governments were all unrivalled. On this occasion he had much less power than the other negotiators, but his ultimate deterrent was the threat to expose their deals and treaties in the courts, and it worked very well.

There were three main points to be discussed: (1) the shareholding for each company; (2) the extent of the concession and the member companies' relationship with each other; and (3) the taxation arrangements.

The question of the shareholding was the most important and provided plenty of scope for inter-company infighting. Deterding and Gulbenkian both remembered with bitterness the way the British Government had secured 50 per cent for Anglo-Persian at the 1914 Foreign Office Conference, and they were determined to reduce its influence. Meanwhile the American group, led by Walter C. Teagle of Jersey Standard, argued that as a point of prestige they could not accept less than Shell or the Compagnie Française des Pétroles. This combined pressure was too much for Anglo-Persian, and in the end it had to accept the principle of equality: each group was given 23¾ per cent of the shares, with the remainder going to Gulbenkian to make up his 5 per cent, and Anglo-Persian was allowed to take 10 per cent of the output for free as compensation.*

The argument over the extent of the concession and the relationship between the member companies was more complicated. In 1914 the Grand Vizir's letter had clearly stated that the concession should cover the vilayets of Baghdad and Mosul, and at the Foreign Office Conference in the same year the companies had

* Reduced to 7½ per cent in 1931.

agreed that they would not hold any interests in the Turkish Empire independently of each other. The American companies pointed out that the empire no longer existed and argued that any such restriction was contrary to their government's policy that there should be an 'open door' into the Middle East for all companies wanting to participate in the area's economic development. They also claimed that any restraint on competition would be bound to bring them into conflict with the authorities in Washington.

In reality the Americans were not so disinterested as they liked to appear. As the Turkish Petroleum Company was primarily a foreign-owned concern, they wanted to restrict its influence as much as they could; they were prepared to join it in order to gain access to valuable oil deposits that would otherwise be out of their reach, but they preferred the idea of a free-for-all in which they could acquire Middle East reserves of their own.

To the French Government and to Gulbenkian the American arguments struck at the very heart of the Turkish Petroleum, and threatened to destroy its entire value. Both wanted the concession to produce the maximum amount of oil over the widest possible area – the French in order to make France independent of the international companies for its supplies and Gulbenkian for the sake of his profits – and they knew that in a free-for-all they would never be able to hold their own. Anglo-Persian and Shell, as large companies that could stand up to anyone, rather favoured the American position. But the French Government threatened to go to law, brought pressure to bear on London, and won the point.

The result was the Red Line Agreement, the most famous example of an arrangement to curtail competition ever made in the international oil industry. It was given this name in October 1927 when the French delegation presented a map on which they had outlined in red the area they considered to have formed the Turkish Empire. The British said it was historically inaccurate, but accepted it just the same, and the Americans followed in their wake. The Red Line encircled virtually the whole of the Arabian Peninsula. Within this area the companies agreed not to compete with each other for concessions nor to hold any individual concessions without first seeking the permission of their partners.

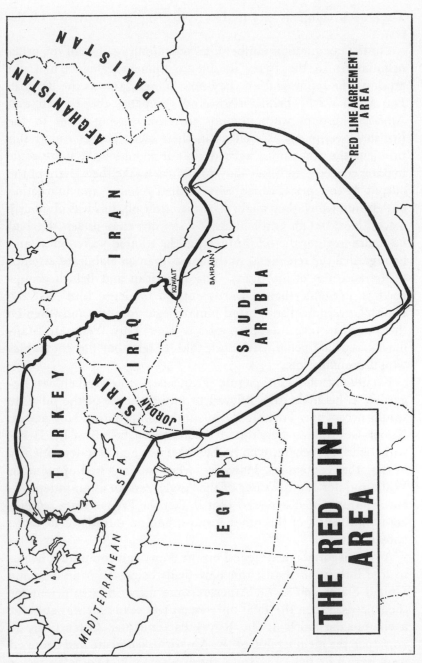

FIG. 4

Thus each company secured a veto over the activities of the rest.

On the tax question Gulbenkian was alone against all the other negotiators. In the 1920s double-taxation agreements had not attained the sophistication they now possess, and as the Turkish Petroleum was a British registered company, the French and American groups were anxious to avoid paying taxes to the British Government as well as to their own. To get round this problem the companies agreed that it should be a non-profit-making concern, and that they would each take their share of the output at cost price. Since they all had refining and marketing outlets into which they could feed the crude oil, this was obviously a good idea, but for Gulbenkian it was worse than useless. He had no desire to enter the oil business; all he wanted was money, and he regarded the scheme as nothing less than an elaborate attempt to defraud him of his rights. The American and British groups tried to persuade their governments to influence him. But the British Government supported him on legal grounds, and when he threatened to take the companies to court Guy Wellman, Standard's associate general counsel, told Teagle that the companies would probably lose.

Eventually the Compagnie Française, who Gulbenkian regarded as his main ally, offered to buy his share of the oil if he would accept the Turkish Petroleum's non-profit-making status. Nubar Gulbenkian negotiated the terms on behalf of his father with Louis Tronchere, who he believed was acting independently of the other companies. However, when he suggested appointing William (later Lord) Fraser of the Anglo-Persian as an independent arbitrator he discovered that in fact the French were secretly acting on behalf of the other groups, and an outsider had to be chosen instead.

As the negotiations dragged on the American companies began to lose interest in them; the new fields being discovered in the United States and Latin America were more than sufficient for their needs. When the final agreement was ready for signature at a champagne lunch at the Royal Palace Hotel, Ostend, on 21 July 1928 the membership of the American syndicate had dropped from seven to five. By 1930 there were only two left: Jersey

Standard and Standard of New York, both of which had extensive markets in Europe and the Far East.

For the Europeans, however, Iraq more than fulfilled its promise. Drilling was delayed for a long time as the details of the concession agreement were being hammered out, but eventually on 30 June 1927 a team 'spudded in' at Baba Gurgur, not far from Kirkuk. Optimism was running high, as the Eternal Fires of Nebuchadnezzar's fiery furnace were only a mile and a half away, and it was thought that the seepages which fuelled them must come from a large deposit. None the less, nobody was prepared for the drama that occurred on 27 October when the drilling bit broke through into the oil reservoir, and a jet so high that it could be seen 12 miles away broke through the well head. For the first time in recorded history the fires had to be put out to prevent an explosion as the oil roared out of the well at the incredible rate of 12,500 tons a day and spread over the surrounding countryside. The flood was eventually brought under control, and when the development drilling programme was completed it became apparent that the field was 60 miles long and one of the largest in the world.

For Gulbenkian, Baba Gurgur was the culmination of thirty-seven years of negotiations, plans, and dreams. Yet he never bothered to go and see it, preferring to remain in Paris surrounded by his works of art and mistresses. It was reward enough to collect the 5 per cent royalty on every ton that came out of the ground, and the traffic jams round the Place de la Concorde provided constant reassurance that sales were going well.

The Iraq Government under King Feisal suddenly found itself the centre of unaccustomed attention. During the complicated negotiations over the concession agreement it had sometimes been rather ignored, and its request for a 20 per cent shareholding in the company was turned down. Its interest in the venture was confined to a royalty of 4s. in gold for every ton produced, but this was more than enough to provide an assurance of far greater wealth than had ever before seemed possible. Moreover, the concession covered only 192 square miles of territory, and after Baba Gurgur the company, which had tactfully changed its name to the Iraq Petroleum Company, naturally wanted more. In return for an additional 32,000 square miles it paid the Government £400,000 in gold, of which only half was to be recoverable from future royalties.

The great irony of the complicated arrangements to develop Iraq's oil is that they were irrelevant to their basic aim. This was to give the United States a stake in the Middle East, and it was thought that Iraq provided the only means of doing so. Yet if anybody had listened to a short, thickset New Zealander called Major Frank Holmes they could have found even larger reserves in Kuwait or Saudi Arabia while the Iraq negotiations were still in progress.

Like Higgins and Lucas, the discoverers of Spindletop, Holmes was not a geologist, and his ideas were scorned by the experts. He was a self-taught mining engineer, and he acquired both his military rank and his interest in oil during the war. He was responsible for buying meat for the British armies in Mesopotamia, which meant travelling widely in the Persian Gulf.* There were plenty of surface signs of oil in the area, such as the bitumen seepages in Bahrain, and what he saw and heard fired his imagination. On his demobilization he returned to London and convinced his former employers, Sir Edmund Davies, Mr Percy Tarbutt, and Mr Edmund Janson to set up the Eastern and General Syndicate to look for concessions and other business opportunities in the Gulf states.

This was a considerable act of faith in Holmes' judgement. Several of the big companies had carried out detailed geological surveys of the region, and despite the surface indications, they were convinced that there were no significant deposits to be found. None the less, they were also anxious not to have outsiders nosing around and putting ideas into the heads of the local rulers. The most powerful of them, by virtue of its monopoly in Persia, was

* The strip of water which the English have always known as the Persian Gulf is claimed by the Arabs to be the Arabian Gulf. Oil companies without interests in Iran have accepted the claim, but any reference to the Arabian Gulf in the Press always provokes a letter of complaint from the Iranian Embassy. Most oilmen now prefer to talk only of the Gulf, but I have retained the traditional English name.

the Anglo-Persian. Its attitude was summed up by its chairman, Sir Charles Greenway, in a minute to the management committee on 9 September 1924, in which he wrote: 'Although the geological information we possess at present does not indicate that there is much hope of finding oil in Bahrain or Kuwait, we are, I take it, all agreed that even if the chance be 100 to 1 we should pursue it rather than let others come into the Persian Gulf, and cause difficulties of one kind and another for us.'

This attitude was echoed by the British Colonial Office, which was responsible for Britain's relations with Sheikhdoms, most of which were in one way or another under British protection. Its policy was to protect the rulers from 'unscrupulous concession hunters', and to keep foreign commercial interests out of the Gulf. Holmes was not, of course, a foreigner, but he was an oilman, and these the Colonial Office regarded with particular suspicion.

The rulers themselves also felt sceptical about the motives of strange men who came to tell them that their deserts contained a black liquid that could be turned into unlimited wealth. The only liquid they were interested in was fresh water, which was more valuable than life itself in the Gulf in those days, and the only wealth they recognized apart from their wives and flocks was gold. Even this, it was felt, could sometimes cause more trouble than it was worth if it led to political upheavals. The relatively prosperous Sheikh Ahmad al Jabir al Sabah of Kuwait, for instance, found the prospect of an oil concession unenticing when set against the complaint of his merchants that it would provide constant employment for the workers and thus lead to a labour shortage in the pearling industry.

Of the places Holmes was interested in only El Hasa in what is now Saudi Arabia was outside British control. It was ruled by Ibn Saud, the great warrior chief who was then in the middle of the series of campaigns by which he united under his rule the Arabian peninsula apart from the British coastal regions. Ibn Saud had a lively distrust of the British, whom he always suspected of trying to subvert his independence. When Holmes and the Anglo-Persian approached him with requests for concessions in 1922 the British High Commissioner, Sir Percy Cox, intervened on Anglo-Persian's behalf. This straight away influenced him in Holmes' favour, and on 6 May 1923 the Eastern and General was granted an exclusive option for oil and mining rights over El Hasa. While his employers hired the eminent Swiss geologist, Dr Arnold Heim, to examine its new properties, Holmes went off to look for new concessions in Kuwait and the island of Bahrain.

In Kuwait he was unsuccessful, but in Bahrain he had better luck. The island was suffering from an acute water shortage. Its main supply came from fresh-water submarine springs on the sea-bed, and severe storms were making it impossible for divers to go down to them. Holmes was staying with a local merchant called Mohammed Yateem, and suggested to him that perhaps artesian wells might provide an alternative source of supply. Yateem passed the idea on to the Ruler, Sheikh Hamad bin Isa, who gave his approval. When the wells struck water Sheikh Hamad was so grateful that as a reward he gave Holmes an exclusive oil option on 2 December 1925.

By now Holmes was much in need of encouragement. He had secured options on what are now two of the most valuable con-cessions in the world, but the syndicate's money was running short, and it was vital to find a buyer among the large companies. Although Heim's report from El Hasa was disappointing, Pro-fessor George Madgwick of Birmingham University was impressed with the geological evidence, and with his backing Eastern and General approached Shell, Anglo-Persian, and Burmah, the three leading British companies. When all three returned negative replies Holmes reluctantly turned to the United States.

There he met with an almost equal lack of enthusiasm. Talks were held with Gulf and Jersey Standard; they regarded Madg-wick's geological reports as interesting but not convincing. More-over, the prospect of mounting an exploration venture in such a distant and isolated spot as Bahrain looked distinctly uninviting, especially in view of the restrictive attitude of the British Govern-ment. Only when the syndicate promised to hand over the rights on the mainland as well as Bahrain could Gulf be persuaded to come to terms, and it was not until 6 November 1927 that the deal was signed, just a week before the Bahrain option was due to lapse.

While Gulf's advance party was preparing to drill, a new and more serious problem arose. Bahrain fell within the Red Line area; in accordance with the terms of the Iraq Petroleum agreement Gulf was obliged to offer its concession to its partners in that group and in the event of their declining to ask their permission to go on alone. On both counts the European companies said 'no'; they neither wanted Bahrain for themselves nor would they allow Gulf to take it.

The syndicate was back where it began, except that it now had Gulf's geologists' reports to confirm those of Madgwick, and the promise of that company's support in its efforts to find another buyer. These were enough to persuade Standard Oil of California, which was not involved in Iraq, to take a chance. Late in 1928 it took up the Bahrain option, even though Eastern and General's rights in El Hasa on the mainland had lapsed through failure to keep up with the rental payments. Another outstanding problem was that of nationality, since the Colonial Office did not want to see Americans working within its sphere of influence, but this was overcome when California Standard agreed to incorporate a special company in Canada to carry out the work.

Within three years Holmes' faith in the Persian Gulf was rewarded when oil was found in Bahrain on 31 May 1932. The major companies in Iraq heard the news with horror. They thought they had secured a monopoly of the Middle East's resources, and now here was a company from outside their charmed circle finding a new and independent source of supply. Moreover, if there was oil in Bahrain there was a good chance that their experts' reports on the rest of the Persian Gulf would turn out to be equally wrong, and a rush for concessions was bound to begin.

Gulf was particularly infuriated over what had happened. After being prevented from being allowed to go into Bahrain it had resigned from the Iraq Petroleum group, and was thus cut off from both the new Middle East fields. It determined to find another, and decided that Kuwait looked the most likely. Accordingly, Holmes, now fifty-two, was sent to the sheikhdom to take charge of the negotiations with Sheikh Ahmad, who was already receiving offers from the Anglo-Persians.

Holmes began the discussions with two great advantages: his reputation as the 'Abu el Naft', the father of oil, was at its height, and he had known Sheikh Ahmad for many years. But there was one point on which he could not match his thirty-year-old adversary from Anglo-Persian, Archie Chisholm: the Sheikh was treaty-bound never to give a concession 'to anyone except a person appointed from the British Government'.* The two men spent over a year in Kuwait making offers and counter-offers, and meeting on Sundays at the small American mission church. While they bargained the British and American Governments argued the case, with the American view being presented by its Ambassador in London, who by a happy coincidence was Andrew Mellon, a former president of Gulf. By the end of 1933 both sides were ready to compromise, and it was agreed to set up a jointly owned Kuwait Oil Company registered in London.

Everybody knew that the concession was highly promising, but not even Holmes had any conception of its true value. He thought it would be no better than Bahrain, and the Eastern and General syndicate sold its right to a small overriding royalty to Gulf for several thousand pounds. In fact, Kuwait is virtually floating on oil and contains about a sixth of the world's proved reserves. Today that over-riding royalty would be worth more than £5m. a year, whereas the Bahrain royalty, which Eastern and General retained, brings in a mere £136,000 a year.

In Saudi Arabia, where Holmes' rights over the El Hasa had lapsed, there were also two main contenders for the concession: the Iraq Petroleum group of major companies, and California Standard. King (as he had now become) Ibn Saud had an instinctive preference for the California company, since it looked to be less powerful than its rival and was unconnected with Britain, but his principal concern was to secure the maximum amount of money. His main source of income was the Hegira, as the pilgrimage to Mecca is called, and because of the world economic depression the number of faithful making the journey was declining rapidly. When the discussions opened the King asked for £50,000 in gold. The Iraq Petroleum group replied with an offer of £30,000 in sterling, but California Standard realized that this was not an occasion for haggling. Within forty-eight hours it had deposited the full quantity of gold, and clinched the contract.

* Federal Trade Commission Report.

CHAPTER 10

Embarras de Richesse

In a sense, the oil industry can never have too much oil. When the basic raw material runs out the companies will lose their livelihood, whereas when there is a surplus they can always try to open up new markets and develop new uses. None the less, the opening up of the Middle East could not have come at a worse moment for the small group of large companies that was trying to establish control over the industry's reserves. In the short run during the early 1930s, the years of the Great Depression, the last thing they wanted was more oil.

Throughout the 1920s a constant stream of new discoveries in the United States pushed prices steadily downwards, and in 1929 the collapse of Wall Street accelerated the decline. In 1930 the demand for oil in the United States fell for the first time, and with Dad Joiner's discovery of the East Texas field in October of that year the bottom fell completely out of the market as prices dropped from $1.30 a barrel to 5 cents a barrel. After years of making large profits the companies suddenly found themselves confronted with enormous losses. In 1931 Shell's American subsidiary alone lost $27m., while Gulf lost $23m., its first ever deficit. The same story was repeated all over the world; in Persia the Government received a 16 per cent royalty on Anglo-Persian's production, and between 1929 and 1931 its value plummeted from £1·4m. a year to £307,000.

In the United States the situation was brought under control through a mixture of state and federal action. As part of his programme for national recovery, President Roosevelt introduced an American Oil Code designed to raise the price of crude by tying it to the price received for petrol during the previous month. For their part the state governments imposed statutory output controls, and when these had been implemented the Federal Government passed the 1934 Connally 'Hot Oil' Act, making it an offence to transport illegally produced oil across state boundaries. It also acted to curtail the amount of foreign oil coming into the country.

In other areas the major companies could take appropriate action on their own account; in Iraq the pace of development was slowed down, and in Venezuela the big three local producers, Shell, Gulf, and Indiana Standard, cut back their output by 15 per cent. The companies would also make arrangements between themselves to prevent a flood of new oil from coming on to the market. This happened twice in Venezuela. The first occasion occurred when the United States imposed its restrictions on imports, thereby cutting off Indiana Standard's Venezuelan fields from their main market. This meant that Indiana would have to find new markets in Europe, where Jersey Standard was strong, and rather than risk having to meet new competition, Jersey stepped in and bought the fields. Later on, when Gulf found a new field, it sold a half interest to Shell and Jersey in preference to looking for new outlets.

In Saudi Arabia and Bahrain the Red Line Agreement made a similar arrangement between the Iraq Petroleum group and California Standard impossible without the consent of the Compagnie Française and Gulbenkian. Anglo-Persian, Shell, and the two American companies, Jersey Standard and Socony Vacuum (formerly Standard of New York), first tried to persuade their partners to remove Saudi Arabia and Bahrain from the Red Line area altogether so that they could buy their way into the new concessions. When they failed they asked to be allowed to purchase California's production; again their partners said 'no'.

The French and Gulbenkian were not simply using their veto to make trouble; the disagreement reflected a basic conflict of interest between the two sides. The big groups saw California's concessions as a threat to their existing interests in the Middle East and the Eastern Hemisphere and were afraid that prices would fall disastrously when the company sought outlets for its oil. The Compagnie Française, by contrast, wanted low prices for the benefit of the French balance of payments, while Gulbenkian wanted the majors to rely as much as possible on the Iraq Petroleum Company in order to maximize the value of his 5 per cent.

California was fully aware of the danger of ruining the market, but could not wait indefinitely for the Iraq group to resolve its internal disputes. Accordingly, in 1936 it sold a half interest in its concessions to the Texas Company, which had an existing Eastern Hemisphere distribution system ready to take the Arabian and Bahrain oil without disrupting prices too much. However, the advantages of a sales agreement with the established majors remained, and in 1939 one was eventually negotiated including the Compagnie Française as well as the British and Americans.

Because of the outbreak of war it never really came into effect, and in any case it would not have satisfied Jersey and Socony Vacuum for long. The Red Line had proved a profitable investment, giving Jersey Standard, for instance, $10 in capital value for ever dollar invested in Iraq. But it had outlived its usefulness, and they were fed up with the limitations it imposed on their freedom of action. By 1941 it was clear that Saudi Arabia's reserves were larger than anybody had previously imagined, and they were determined to buy their way into that country.

The war gave them their opportunity by disrupting all the Iraq Petroleum arrangements. With the fall of France the Compagnie Française and Gulbenkian, who remained in Vichy,* were declared enemy aliens, and cut off from their oil, while even the allied groups could not carry on normally. Jersey Standard and Socony Vacuum claimed that as a result the Red Line Agreement had been dissolved, and that to reinstate it would be contrary to the American anti-trust laws. Without waiting for the consent of the French or Gulbenkian they began working out the terms of a new deal with California and Texas, and in March 1947 Jersey received a 30 per cent share in Aramco, as the Saudi production company is called, in return for $76·5m., while Socony Vacuum got 10 per cent for $25·5m.

The Aramco purchase meant that the vast Saudi Arabian reserves could be absorbed into international markets through existing distribution systems, thereby saving California and Texas

---

* His son Nubar stayed in London, thereby giving the family a foot in both camps.

a good deal of capital investment, and simultaneously protecting Jersey and Socony Mobil from a new source of competition. A similar deal was also needed for Kuwait, where Anglo-Iranian (formerly Anglo-Persian) and Gulf were sitting on large reserves without having the necessary marketing facilities. Jersey Standard and Socony Vacuum therefore signed long-term contracts to buy oil from Anglo-Iranian, while Shell did the same with Gulf. Thus by the end of 1947 the interests of all the major Anglo-American groups in the Middle East had been brought into a happy state of harmonization.

The Compagnie Française and Gulbenkian objected violently, but the French were in a weak bargaining position. France desperately needed oil to repair its war-damaged economy, and Iraq was the only country where it had a stake of its own. The French were afraid that if they held out against the big groups these would delay the development of the Iraqi fields. Accordingly, when the Compagnie Française was offered a guarantee of rapid expansion in Iraq and told that it could take more than its *pro rata* share of the output in return for agreeing to the end of the Red Line it had no choice but to accept.

Gulbenkian, living as a recluse in the Hotel Aviz in Lisbon, was more intransigent. He hired Sir Cyril (now Lord) Radcliffe to defend his interests, and began legal proceedings in London. Once again the threat of exposure in open court proved a powerful weapon; at two o'clock on the morning of the day his case was due to be brought forward the companies agreed to his terms. The Red Line Agreement was formally dissolved with effect from November 1948, and as compensation Gulbenkian received an extra 3·8m. tons of oil a year for fourteen years in addition to his 5 per cent.

Although Calouste Gulbenkian died in July 1955, his 5 per cent lives on. It must be one of the most profitable investments of all time. In 1965 his son Nubar estimated that from 1914 to 1953 the Gulbenkian interests never had more than £1m. at stake in the Middle East at any one time, and that since 1955 alone they have been receiving a return of between £5m. and £6m. a year.*

* *Pantaraxia*, Nubar Gulbenkian. Hutchinson.

CHAPTER 11

The attempts to form a cartel

The major companies' efforts to establish a monopoly – or rather an oligopoly – over the production of oil were not surprisingly accompanied by a similar attempt to control its distribution. The aim was to establish an international cartel to regulate prices and competition in such a way as to guarantee a profit to all the members. All the leading companies were involved at one time or another, but the most active were Jersey Standard, Shell, and Anglo-Persian, the three with the most widespread interests. Shell and Jersey Standard had, of course, been operating on a large scale before 1914, and Anglo-Persian came into the reckoning immediately the war finished. During the four years of fighting the production from its Persian fields grew rapidly, and with the arrival of peace it had built up a distribution system to take the place of the armed forces. The British Government helped by presenting it with various German-owned oil interests* seized as alien property.

The move towards a cartel did not begin immediately. At first the companies were far too busy meeting the rapid growth in demand, which more than doubled between 1919 and 1926,† and in building up a modern sales network. The pattern was set in the United States, where the first drive-in filling station is believed to have been opened in Pittsburgh by Gulf in 1913. By 1920 similar stations were to be found throughout America, and by 1929 there were more than 29,000 in Britain alone. In some places,

* One of these was the British Petroleum Company, so called because it sold refined products in Britain.

† From 568·6m. barrels to 1,173m. barrels. In 1919 the United States used 344·5m. barrels and in 1926 780·5m. Non-American demand (excluding Soviet Russia) rose from 194·1m. to 393·1m.

such as Britain, they were usually owned by independent dealers, while in others they were owned by the companies themselves.

In several countries the companies had price-fixing agreements, but these were not taken very seriously. Whenever one of the parties thought it could gain an advantage by starting a price war it would do so, and the deal would go by the board. In Holland, for instance, there was an agreement among the leading companies not to undercut each other for the business of established customers. It guaranteed Jersey's local affiliate nearly 80 per cent of the market. Deterding regarded such a situation in his home country as outrageous, and in 1924 Shell launched a price war in an attempt to reverse the position.

Elsewhere the competition could be even tougher. In 1922 Jersey was horrified to learn that a state monopoly on the importation and distribution of oil had been imposed in Iceland, a country it had supplied for thirty years. Its local manager could not understand what had happened, but the reason for the decision soon became clear when the Government signed a three-year exclusive supply agreement with Anglo-Persian.

In general, however, competition was kept under control until September 1927, when a price war broke out between Shell and Standard of New York (the company which in 1931 changes its name to Socony Vacuum). It started over a dispute about Russian oil, a subject on which Deterding felt strongly. His wife Lydia Pavlovna – the woman to whom he once gave the £300,000 parure of emeralds before their marriage – was an exiled Russian aristocrat, and not unnaturally loathed the Communists. Deterding also had good reasons of his own for not liking them, since they had seized all Shell's production facilities in the Caucasus. When at last all hope of receiving compensation disappeared Deterding demanded that Standard of New York should stop buying Russian oil for sale in India, and when the Americans refused he tried to force them out of the market by starting a price war.

What began as a purely local struggle quickly spread to Europe and the United States. More and more countries were dragged in, and the other companies began to suffer through being forced to cut their prices in order to retain their positions. The fight could not continue indefinitely, and eventually at the end of 1928

Shell, Standard of New York, and Anglo-Persian reached an agreement on how the Indian market should be shared out. By itself this would not have been particularly important, but it became the first step in a much wider settlement.

All the international companies had been thoroughly frightened by the way an apparently local dispute had spread across the world and felt that it could not be allowed to happen again. The finishing touches were just being put to the concession agreement in Iraq, and the companies decided that closer co-operation was equally necessary at the marketing end of their business.

Outside the United States, where there was an almost infinite number of small companies and stringent anti-trust laws, marketing was already in sufficiently few hands to make a cartel seem a reasonable objective. Just over 50 per cent of the total sales were controlled by the big three – 23 per cent for Jersey, 16 per cent for Shell, and 11·5 per cent for Anglo-Persian. The Russians had about 6·5 per cent, and most of the remaining 43 per cent belonged to large American companies such as Standard of New York and Texas. The number of real outsiders looked just about as significant as a Balkan army when Russia and Germany are at war.

Throughout the summer of 1928 secret conferences sponsored by the three majors took place. When at last everything was ready Deterding invited Walter Teagle, the chairman of Jersey Standard, and Sir John (later Lord) Cadman,* the head of Anglo-Persian, to Achnacarry House in the Scottish highlands, a hunting lodge belonging to The Cameron of Lochiel, which he had ostensibly hired for the grouse shooting. The house is situated in a wild stretch of moorland country, and the scene when the three most powerful men in the oil world gathered there could have been taken straight out of the pages of one of those John Buchan thrillers in which powerful anonymous men would spend weekends at Scottish castles planning great *coups* for the destruction of Britain's enemies.

Deterding was the dominant figure at the meeting. For nearly thirty years he had been among the leaders of the international

---

* The same man who in 1913 had headed the Cadman Commission of inquiry into Persian oil for Churchill before the Government decided to buy its way into Anglo-Persian.

oil industry, and under his direction Shell had enjoyed a more spectacular rate of growth than any other company. Ever since his early days with the Royal Dutch he had preached that, 'Co-operation means strength'. Although he could cut prices with the best, he had never changed his opinion that stable prices and a close understanding between the major companies was the best way to achieve higher sales and bigger profits. No doubt both Deterding and Teagle remembered the occasion in 1907 when Deterding expressed these views to John D. Archbold, who was then the executive head of the whole Standard empire. They had been quickly rejected, but now the two men had taken over from Archbold's generation and could impose their own ideas on the industry.

The result of the Achnacarry conference was an agreement to set up an international oil cartel. Its official title was the Pool Association of 17 September 1928, but it is usually referred to more simply as the Achnacarry Agreement. It covered the whole world outside the United States and the Soviet Union, and within a few months the grouse party had persuaded a large number of other American companies* with big overseas interests to accept its main provisions. These were clearly laid out in a list of seven principles, by which it was hoped the signatories would in future guide their conduct. As these constitute a sort of Declaration of Intent, which was to form the basis of several efforts to control the world market during the 1930s, they are given here in full:

1. The acceptance by the units of their present volume of business and their proportion of any future increases in consumption.

2. As existing facilities are amply sufficient to meet the present consumption, these should be made available to producers on terms which shall be based on the principle of paying for the use of these facilities an amount which shall be less than that which it would have cost the producer had he created these facilities for his exclusive use, but not less than the demonstrated cost to the owner of the facilities.

---

* Atlantic Refining, Cities Service, Continental, Gulf Maryland, Pure, Richfield, Sinclair, Standard of California, Standard of Indiana, Standard of New York, Texas, Tidewater, Union, Vacuum.

3. Only such facilities to be added as are necessary to supply the public with its increased requirements of petroleum products in the most efficient manner. The procedure now prevailing of producers duplicating facilities to enable them to offer their own products regardless of the fact that such duplication is neither necessary to supply consumption nor creates an increase in consumption should be abandoned.

4. Production shall retain the advantage of its geographical situation, it being recognized that the value of the basic product of uniform specifications are the same at all points of origin or shipment and that this gives to each producing area an advantage in supplying consumption in the territory geographically tributary thereto which should be retained by the production in that area.

5. With the object of securing maximum efficiency and economy in transportation, supplies shall be drawn from the nearest producing area.

6. To the extent that the production is in excess of the consumption in its geographical area, then such excess becomes surplus production, which can only be dealt with in one of two ways: either the producer to shut in such surplus production or offer it at a price which will make it competitive with production from another geographical area.

7. The best interests of the public as well as the petroleum industry will be served through the discouragement of the adoption of such measures the effect of which would be materially to increase costs, with the consequent reduction in consumption.

The principles speak for themselves. In an effort to stem the rising tide of competition the big three agreed to try to freeze the market in its existing mould. They were to combine their interests and share each other's facilities – refineries, storage, tankers, and the rest – in order to present a united front against companies trying to break into new markets, price cutters, and other disturbing elements. In this way they hoped to derive the maximum benefit from their control of the new oilfields in Iraq and Venezuela, and of their other fields spread across the world; each field would supply the markets nearest to it, thereby saving the cost of unnecessarily long tanker voyages. If a company did not happen to have a share in the field nearest one of its markets there was no need to worry: an exchange agreement could easily be worked out with one who had.

In essence therefore the companies were agreeing to function virtually as joint ventures, with each shareholder contributing a certain proportion of oil and capital investment and receiving a share of the profits in exchange. But it was not an exclusive cartel. The big three realized that a closed shop is certain to fail, since the outsiders can then pick up as many customers as they want by slightly undercutting the cartel price, so that after a while the members are forced to break ranks to maintain their share of the sales. The majors' aim was to reach agreement with all the leading operators in every market, and particularly the other large American companies.

Naturally all this involved working out a system of quotas for each market and agreeing on a formula for fixing prices. The distribution of quotas, the allocation of transport, the fixing of freight rates, and various other administrative matters were made the responsibility of a central association in which each member was represented. At the same time two American export associations were set up to allocate the total American quota among the various companies involved. One called the Standard Oil Export Corp. was formed in December 1928 and consisted of Jersey Standard and five of its subsidiaries, while the other, the Export Petroleum Association Inc., formed early in 1929 included the Standard corporation plus the other big American exporters. Standard thus acted as the link between the American industry and its European rivals.

The problem of prices was solved by the formulation of the Gulf-plus pricing system. This laid down that the price of oil should be the same in every export centre throughout the world as in the American ports along the Gulf of Mexico. But the final cost at the point of delivery should vary depending on its distance from the Gulf of Mexico and on whether or not the buyer was a member of the cartel.

Ordinary commercial customers and oil companies outside the cartel had to pay the basic Gulf price plus the cost of shipping the oil from the United States to the point of delivery, wherever that happened to be. It made no difference to the buyer where the oil actually came from, and if he could be supplied from a field nearer than the Gulf of Mexico all the savings went to the cartel company. If a buyer in Bombay, for example, placed an order with a cartel company the oil would probably be supplied from nearby Persia or Iraq, but he was still charged the same freight cost as if the oil had been brought all the way from the United States.

When companies within the cartel were selling to each other profiteering on this scale could not be allowed. The Gulf of Mexico price was still used as the peg for pricing all over the world, but any freight savings achieved by drawing supplies from another source were shared equally between buyer and seller. In the nature of things, there was a good deal of buying and selling within the cartel, since the members had promised to share each other's facilities, including oilfields, and, when possible, to supply their markets from whichever field happened to be nearest. Thus if Anglo-Persian, for instance, needed 100,000 tons of oil for Britain it would look round for a company with fields in the United States, since the distance from the United States to Britain is less than from the Anglo-Persian's own fields in the Middle East. If it found a seller among the cartel companies the two of them would then share the saving in transportation costs. On other occasions the cartel companies would go in for straightforward swaps; Anglo-Persian might buy oil from Shell's American fields for delivery to Britain, and in return supply Shell's distribution facilities in Italy with the same amount from the Middle East.

The Gulf-plus system did not represent an entirely new concept. For many years the companies had based their prices on those of the United States, and it was natural that they should. With the disruption of the Russian industry during the Communist revolution the United States emerged from the war with 70 per cent of the world's production, and by 1938 it still accounted for 68 per cent. In any market a country accounting for such a high proportion of the total output would be used as the basing point for international prices. This does not, of course, explain why a customer living near an oilfield, such as the buyer in Bombay drawing supplies from the Persian Gulf, should have to pay the same price for his oil as if it had come all the way from the

United States. He should have been given the benefit of the much lower cost of bringing oil from the Persian Gulf. The explanation lies in the fact that from whichever country a buyer's oil was drawn it was likely to be supplied by a major company, and as a group these companies' largest and most valuable investments lay in the United States. It was therefore in their interest to maintain the largest possible market for American oil, and if that meant charging a high price for Middle East and Venezuelan supplies, then so much the better for profit margins.

If the new fields in the Middle East and Venezuela had been discovered by companies without any assets in the United States the situation would have been quite different. There would have been a repetition of what happened in the 1880s when Russian oil flooded into Europe to break Standard's domination, and forced the company to reduce the price of its American exports. Production costs in the Middle East and Venezuela were about half those of the United States, but Middle East and Venezuelan oil none the less always sold at the same prices as if it had come from the United States. To the majors the idea of a price war between their new fields overseas and their old ones in the United States was absurd. Far better for them to fix prices designed to secure a profit on all their operations, and to hold back production in the new fields. Only Anglo-Persian had no investments in the United States; and it was not only too small to take on the others alone, it also found the Gulf-plus system extremely profitable.

The maximization of profits was not the big companies' only aim. Their executives (like the governments of oil-producing countries today) regarded oil as a scarce resource to be husbanded rather than thrown on to the market at a low price just because competition could not be kept within bounds.

Their main effort to form a cartel occurred outside the United States because the concentration of sales in the hands of a small number of companies made the objective seem easier than in the United States, where the industry was very fragmented and there were strict anti-trust laws. But even in the United States the big companies tried to support prices by reducing competition. The task was undertaken by the American Petroleum Institute, and on 15 March 1929 it unveiled a far-reaching plan for production controls at a meeting in Houston. However, the Attorney-General refused to give it his blessing, and it was not until after the discovery of the East Texas field led to the complete collapse of the market during the Depression that the Government was converted to the idea of output controls and the conservation of resources.

Soon after the failure of the Institute's initiative the American end of the international cartel also ran into trouble. The Export Petroleum Association could not secure the co-operation of the small independent companies, who continued to export their oil at lower prices than the cartel members. At the same time the members themselves found they could not agree on the level at which the Gulf price should be fixed. Shell's American subsidiary wanted it to remain fairly low, while the Americans thought it ought to rise. As decisions could be taken only by a unanimous vote, an impasse was reached, and by November 1929 the association had collapsed.

Although the cartel had proved more difficult to organize in practice than to work out on paper, the big three were not unduly depressed. Despite the overall failure, good progress had been made in several areas. In Britain, the second largest market, things had gone particularly well with the conclusion of a series of agreements that promised drastically to reduce competition if not to eliminate it altogether.

Britain was especially suitable for cartel operations, since over 80 per cent of the market was in the hands of the 'national companies', as Shell, Anglo-Persian, and Jersey Standard's subsidiary Anglo-American were usually called. For some time before Achnacarry they had had a price-fixing agreement among themselves, and observed certain other rules designed to ensure 'orderly' competition. But a number of independent companies, including the Soviet-owned Russian Oil Products, continued to cut prices, and generally make a nuisance of themselves. However, by the agreement of 1 January 1929, everything was put on to a new footing. The Big Three undertook to give all dealers the same profit margin instead of discriminating against those selling independent products, while the independents promised to adopt the big companies' price schedules, and to sell only to 'legitimate'

* See Chapter 10.

customers instead of looking for new outlets. At first the Russians held out, but within a few months they, too, joined in.

Under these circumstances the wiser heads began to see that perhaps Achnacarry had been too ambitious. As Mr E. J. Sadler, a vice-president of Jersey Standard, put it: 'The making of a world-wide agreement is more difficult to obtain than accomplishing the result piecemeal. Economically there are local situations which can be consolidated with a much sounder economic basis than to immediately attempt to jump to a position of world-wide distribution.'* Between 1930 and 1934 continuous efforts were made to proceed along these lines, and three separate agreements were signed – the 1930 Memorandum for European Markets, the December 1932 Heads of Agreement for Distribution, and the June 1934 Draft Memorandum of Principles. All were basically restatements of the original Achnacarry principles designed to take account of the practical problems that had arisen when they were put into practice, and none of them was completely successful. Even the Draft Memorandum, which remained in effect until the war, was treated rather like the Ten Commandments, as a set of rules to which all pay lip service but few follow to the letter.

The first three arrangements were all overpowered by the effects of the Great Depression. In 1931 world oil consumption declined for the first time in the twentieth century; in 1932 it fell again, and not until 1934 was the 1930 level recovered. Simultaneously more and more oil was becoming available from the new discoveries in Texas and the Middle East. Every company† was suffering severe losses, and whatever agreements might be reached in principle, it was impossible to prevent competition from breaking out. The pressure of supplies was such that if an opportunity arose for a marketing manager to steal another company's customer the temptation was too strong to resist, regardless of the chairman's promises. Oil companies, by this time, had grown into vast concerns with subsidiaries in many different countries, and it was impossible for the head offices to follow every detail of what was happening. In any case, in commercial life a balance sheet and a sales graph usually carry more weight than a promise to a competitor. Only the 1934 Draft Memorandum

* Federal Trade Commission Report.   † See Chapter 10.

started life with any chance of success. It was launched after the worst of the depression was over, and when the United States Federal and state governments' measures to restrict oil production and raise prices were beginning to take effect.*

The Draft Memorandum was drawn up in great secrecy at a meeting in London in April and May 1934 attended by representatives of Jersey Standard, Shell, and Anglo-Persian, with Jersey acting as representative for the American industry. Even by the exacting standards of oil-company diplomacy, a strong emphasis was placed on secrecy. Only the most senior officials were allowed to know all the plan's details, and they were warned that 'it is important that you realize that this memorandum is to be treated with the utmost confidence, and should not be copied or circulated except possibly to the responsible heads of departments affected by the arrangements'.† It was designed to cover all countries outside North America. A committee was established in London to run the operation, and it was planned to establish separate agreements for each particular market; to ensure their effectiveness the participants hoped to persuade as many companies as possible to co-operate.

The scope of the Big Three's commitments to each other was wider than in any previous agreement. In addition to the usual rules concerning the allocation of quotas and the penalties to be imposed on those who either sold too much or too little there was an agreement to consult on such intimate matters as capital investment and advertising. In the words of Clause 21: 'It is agreed that budgets covering certain items of capital expense and advertising budgets should be, insofar as possible, agreed upon for each market before being submitted for consideration to London with a view to eliminating unnecessary duplication and thus placing an unnecessary burden on the consumer.'‡

In theory, co-operation was to be on a grand scale, but in practice it varied considerably from place to place. In Britain Shell and Anglo-Persian§ had amalgamated their distribution organizations in 1932, and the new joint company Shell-Mex and

---

* See Chapter 10.   † Federal Trade Commission Report.   ‡ Ibid.

§ The two companies made similar amalgamations in several other countries as well.

BP maintained a close relationship with Jersey Standard's sub-sidiary Anglo-American until the imposition of the wartime controls. Throughout this period they continued to control over 80 per cent of the market in all the main refined products, and with the smaller companies adhering to cartel prices, competition virtually disappeared except when the retailers themselves got involved in private feuds.

Sweden was another country where co-operation was particularly close, and it also provides the best-documented evidence of the cartel's activities as a result of the 1947 report by the Riksdag's Oil Investigating Committee. When the local cartel discussions began in 1930 some 97 per cent of the market was in the hands of six companies – the ubiquitous Big Three plus Texas, the small local firm of Nynas, and the Russian-owned Nafta. In addition, Gulf controlled another local company called Alfred Olsen, and in 1937 it bought out the Russians. The first four led the way in evolving a scheme for fixing prices, classifying customers, and regulating competition, but all the others, including Nafta, quickly joined in.

The administration of the cartel was handled by a weekly meeting of senior officials at the Shell offices, and their minutes show how close their co-operation could be. On 31 January 1936, for instance, it was agreed that the Stockholm tramways was getting its petrol too cheaply from Alfred Olsen, and that the price should be raised; just in case the tramways looked for an alternative supplier, the other companies promised that if asked they would quote an even higher price. The following year the same tactics were repeated when Shell, which supplied the tramways with gas oil, raised its price by 20 per cent after receiving assurances that nobody else would undercut it. Yet even in Sweden there were too many companies involved to make a perfect cartel possible. Whatever the senior officials agreed at their weekly meetings, it was difficult to persuade the junior managers from sometimes taking the law into their own hands, and the managing director of Shell later told the investigating committee that no more than 50–60 per cent co-operation was ever actually attained.

Elsewhere things were more difficult, and some of the factors that had ruined the first three efforts to achieve a cartel continued to cause trouble. For all their power, the majors could never gain complete control over most of the larger markets, any more than Rockefeller had been able to impose his will in the United States during the nineteenth century. In most places the three major companies found at least three or four other companies had to be taken into account, and it was not always possible to reconcile their interests.

It was also difficult for the large companies to harmonize the policies of their subsidiaries in different countries. Oil is not easy to store, and when a sales organization finds that it cannot sell all it has in its own country it looks for outlets overseas.

In 1936 my father, Dr Georg Tugendhat, and Dr Franz Kind started an independent refining company in Britain called Manchester Oil Refinery. This was contrary to the interests of the cartel, and a leading figure in Shell warned them that they would not be able to buy any supplies. However, without much difficulty they found an American broker, who dealt with oil on a wholesale basis, and he provided them with cargoes purchased from the Shell subsidiary in the United States. The major companies also tried to prevent Manchester Oil Refinery from selling its output in Britain, and this problem was circumvented when the Gulf subsidiary in Belgium agreed to buy it.

Another problem facing the cartel was that governments were becoming increasingly involved in oil affairs. Coal, steel, and the railways, the traditional commanding heights of the economy in industrialized countries, had always been subject to close official control. It was therefore inevitable that as oil became more important it would be subjected to the same discipline.

Among the larger countries the most active was France. In 1928 it introduced a quota system to regulate each company's imports, and in 1931 it started to encourage the building of refineries so that the companies would import crude oil instead of the more expensive refined products, thereby saving foreign exchange. At the same time it tried to help the Compagnie Française to dispose of its Iraqi crude by establishing the Compagnie Française de Raffinage.* This new concern was then

* The Compagnie Française de Raffinage was 55 per cent owned by the Compagnie Française des Pétroles, 35 per cent by various French independents, and 10 per cent by the Government.

given the privilege of refining 25 per cent of all France's requirements. It was supplied by the Compagnie Française, and the international companies had to buy its products regardless of whether they wanted to or not.

The Italian Government under Mussolini tended to follow much the same line as the French, only with less vigour. In 1926 it established the state-owned Azienda Generale Italiana Petroli (commonly called Agip), which looked for oil both in Italy and the Middle East. In Iraq it secured a share in a small concession, but when Italy ran short of foreign exchange during the war against Ethiopia Mussolini sold it to pay the dues demanded by the Suez Canal Company on the ships supplying his armies in Africa. Nevertheless, the Italian effort was not entirely in vain. The state railways found some low-grade crude oil in Albania, while Agip was able to take over some of the Rumanian producers. In 1932 Italian refiners were given a preferential position in the domestic market through the imposition of a tax on imported refined products, and by 1939 Agip accounted for about a quarter of the local sales.

Germany was luckier than its neighbours in having a few small oilfields of its own in the north-western part of the country, largely owned by German companies. Because of oil's importance to the Luftwaffe, these came under the personal care of Hermann Goering, and in general they were left much freer to carry on their own affairs than many other sections of the country's economy. But there was never any doubt that the interests of the Government had to be put before anything else, and the local affiliates of the major companies were compelled by law to become responsible to the State instead of to their shareholders.

In most other parts of the world some attempt was made to give local capital a stake in the industry. The Japanese Government, helped by the existence of some small reserves in its own home islands, encouraged the formation of locally owned refining and distribution companies; in Argentina the State Oilfields Department (YPF) was given a privileged position in the market and the power to allocate import quotas; in Chile a private company strongly backed by the Government was given a 20 per cent share of all the petrol sales; while in Spain in 1928 a government-sponsored organization was given a monopoly of all distribution.

Yet despite the Governments, the continued existence of the independents, and their own failures the majors did better out of the cartel arrangements than they would have done otherwise. The Gulf-plus price structure remained the basis on which the vast bulk of the world's oil was priced regardless of its place of origin, and this established the framework within which most competition took place. Under these circumstances any company with fields all over the world, and not just the Big Three, was bound to come out on top in the end.

Deterding had the satisfaction of seeing the ideas about stable prices and co-operation between companies, which he had outlined to Archbold as a young man, put into effect. When he died in February 1939 he had good reason to be satisfied with their success.

CHAPTER 12

Oil and the Second World War

'Modern warfare depends on armament and armament depends on oil.'

Ralph K. Davies

At the outbreak of the Second World War in September 1939,
Germany found itself in a similar position to 1914. Through the
major oil companies the United States, Britain, Holland, and
France controlled over 80 per cent of the world's supplies, and the
Royal Navy was strong enough to prevent much of it from reaching Germany. Of the remainder, by far the most important source
was the Soviet Union, and the only other significant producer in
Europe was Rumania. Together these two countries accounted for
about 13 per cent of world production.

The main oil-producing countries in 1938

	Tons
United States	161·9m.
Venezuela	27·7m.
Iran	10·2m.
Mexico	5·5m.
Iraq	4·4m.

Oil was the Achilles heel in the German economy as it prepared
for war. In 1938 the Reich imported nearly 41m. barrels, mainly
via Jersey Standard, Shell, and Anglo-Iranian, while its indigenous production amounted to a mere 4m. barrels. These figures
overstated the country's weakness, since the German producers
wisely took the view that the best way to store their oil was to keep
it in the ground, and there was plenty of scope for a dramatic

increase in output once the war began; moreover, much of the imported oil went straight into the national stockpile, which at the time of the invasion of Poland was estimated to hold about 50m. barrels.

Nevertheless, Hitler realized that for as long as the Royal Navy controlled the sea routes to Europe Germany would never be able to get enough oil for its needs, and a major programme was launched for the development of a synthetic industry based on coal.

The synthetic products were very expensive to make, and not as good as the real thing. This was especially true of petrol, and in the Battle of Britain the Spitfires owed much of their superiority to the fact that they could use 100-octane fuel, whereas the engines of their German adversaries had to be designed for use with the lower-quality synthetic fuel, which reduced their acceleration and climbing powers. However, synthetic oil was a good deal better than none at all, and under the auspices of the Wirtschaftsgruppe Kraftstoffindustrie twenty-five plants were brought into full operation by 1939, with more under construction; by 1941 their total output was over 30m. barrels a year.

Germany's oil shortage was one of the factors behind the General Staff's decision to stake everything on the blitzkrieg method of attack. The generals remembered Lord Curzon's remark that in the First World War 'the allies had floated to victory on a wave of oil', and they also remembered Ludendorff's* complaint that lack of oil had played an important role in bringing Germany to its knees. They knew that despite its enormous industrial and military strength, the Reich could not afford another prolonged static war on a wide front, and that only by accumulating stocks of oil and other essential raw materials and then making a sudden lightning attack on carefully selected points could they hope for victory. This helps to explain both the lull between the Battle of Poland and the Battle of France, known in England as the 'phoney war', and the quiet weeks in between the Luftwaffe's mass attacks on London and other British cities.

* General Erich Ludendorff, first Quarter-Master General and *de facto* Commander-in-Chief of the field army of the German Empire from 1916 to 1918.

The German strategy worked brilliantly. First Poland and then Western Europe fell like a pack of cards, and at the end of 1940 Germany found itself with more oil than when the campaigns began. The Wehrmacht and Luftwaffe managed to win their victories using a mere 12m. barrels of oil products, or about the same as the United States produced every three days, and in France and the other occupied countries they were able to take over stocks amounting to twice as much.

If Britain had fallen as easily Hitler's problems would have been solved. There would have been no need to occupy the British Isles; an honourable peace and the lifting of the blockade would have been enough. As it was, Britain's decision to hold out forced Germany into precisely the kind of long-drawn-out war of attrition that the General Staff had hoped to avoid.

The very success of his armies on the Continent made Hitler's task against Britain more difficult; for by taking over so many countries he made himself responsible for their problems, and the most important of these was the inability of their economies to function properly without imports. It was all very well to re-organize Europe into a New Order, and to give the conquered countries the job of providing for Germany's needs. But neither Germany nor the rest of Europe could play their allotted role unless their economies were supplied with fuel, food, and raw materials, and these Hitler could not provide so long as the Royal Navy controlled the ocean trade routes.

Oil was among the most urgent requirements, and its shortage revealed his problems in their starkest form. Under normal circumstances Poland and Hungary had just about enough domestic production to cover their own modest requirements, and Albania's small fields, taken over by Italy, were another source of supply. But even after the consumption of the occupied countries had been reduced to less than a third of their peace-time level, and severe rationing had been imposed on Germany and Italy, there was not enough to go round. Hitler, like Napoleon before him when encircled by a British blockade, was forced to extend the area of conflict. In October 1940 he gained control of Rumania's oilfields, whose production amounted to 43m. barrels a year, the equivalent of Germany's pre-war annual imports. But transporting the oil from Rumania to Germany was not easy, and the shortage continued. Only two possible alternatives remained: the Middle East, which the British were defending in North Africa, and the Soviet Union.

Under the terms of the Molotov–Ribbentrop pact the Russians were already supplying large quantities of oil and other industrial raw materials* to Germany, and between January 1940 and June 1941 they delivered some 16m. barrels of oil. None the less by the spring of 1941 there was a growing feeling in Berlin that the German economy needed even larger contributions. On 23 May 1941 Hitler sent Stalin a secret memorandum calling for the joint exploitation of Russia's oilfields, and Stalin's refusal was among the immediate factors which led to the invasion of Russia. Obviously there were other considerations which caused Hitler to embark on this disastrous adventure; the decision must be seen in the context of his fear of Communism and his desire for world conquest. Nevertheless, there is no doubt that the desire for more oil played an important part in his calculations, and it is significant that the spearhead of the German advance was directed towards the Caucasian oilfields rather than Moscow.

If the German armies had been able to take over the Russian fields intact and to maintain their communications with the West, Germany's oil problems would have been solved. In 1940 Russia produced 213m. barrels, enough to supply the full 1938 needs of Germany and the countries under its control, and the German engineers could have stepped this up a good deal further. Thus when the invasion failed, Hitler suffered an economic setback which was overshadowed only by the extent of the military disaster.†

On the other side of the world Japan faced many of the same problems as Germany; by far the strongest military power in Asia, it lacked the rice and industrial raw materials, such as rubber and oil, which were vital to its ambitions. After the fall of France it was able to acquire virtual control over Indo-China, which helped to alleviate the rice problem, but the economy

---

* Including grain, cotton, timber, manganese, and chrome.

† The best account of oil's role in the war up to this point is contained in 'The Paradox of Oil and War', Walter Levy, published in *Fortune* in 1941.

remained dependent for oil on imports from the international companies. In July 1941 these were cut off when the United States, followed by Britain and Holland, imposed a ban on all normal trade with Japan. This action was taken in response to the Japanese armies' threats to American, British, and Dutch interests in Asia, but its effect was unfortunately to strengthen the forces of aggression rather than those of peace. In the view of Harold Macmillan: 'It was the ban on oil which may well have persuaded the Japanese Navy, generally regarded as favourable to moderate policies, to agree to take the plunge.'*

As soon as the decision to embark on war had been taken, the principal strategic aim of the Japanese became the capture of the oilfields in the Dutch East Indies and Burma. Notwithstanding their tremendous successes following the surprise attacks on Singapore and Pearl Harbor on 7 December 1941, they failed to achieve this objective. In the East Indies the Dutch destroyed 88 per cent of the production facilities worth over $500m., while in Burma the British blew up as much as they could before the enemy arrived. Some of the facilities were eventually brought back into service, but Japan, like Germany, was forced to rely increasingly on synthetics as the war went on.

Although the loss of Singapore and the military defeats in South-East Asia were major military humiliations for the British, the fact that the Japanese attacks had at last brought the Americans into the war more than made up for the losses. The United States has been described as being 'like a gigantic boiler. Once the fire is lighted under it there is no limit to the power it can generate.' For Britain in December 1941 this power meant not only armies of almost unlimited size but also access to America's raw materials and the use of its ships.

In oil, as in everything else, this brought about a dramatic improvement in the country's chances of achieving eventual victory. Even before the United States entered the war, Britain, with its control of the sea and possession of oilfields all over the world, was, of course, far better placed than Germany. The problem was to get the oil from the ports of North America and the Middle East to the British Isles, and from September 1939

* *The Blast of War*, Harold Macmillan. Macmillan.

5. Signal Hill oilfield, California. The picture shows how closely the Americans used to drill their wells

6. (*a*) Leonard F. McCollum

6. (*b*) Enrico Mattei

6. (*c*) J. Paul Getty

onwards the German Admiralty concentrated the full venom of the U-boat attacks on tankers.

The policy worked with devastating effect. In the first six months Britain lost twenty-one tankers, or more than 5 per cent of its fleet,* and by July 1941 more than half its ocean-going capacity had been sunk. This was not as bad as it might have been, since large numbers of ships from the Norwegian, Dutch, and other European fleets entered the British service when their countries were overrun by the Germans, and the Americans helped by allocating fifty tankers to carry supplies from the Caribbean to Halifax on the Canadian coast, thereby shortening the transatlantic journey by over 1,000 miles. Nevertheless, by the end of 1941 the problem of maintaining supplies was becoming increasingly serious.

When the United States entered the war the tankers carrying oil from Venezuela and the Gulf of Mexico to New York and the north-eastern states lost their immunity, and the U-boats stepped up their attacks. For a while they were even more successful than before, and in the winter of 1942–43 it looked as if they might succeed in cutting the New England states off from the petrol and heating oil on which their people had come to depend. Demonstrations and riots broke out in several areas, with the mobs over-turning fuel trucks to get at their contents, and Americans began to learn what rationing meant. But the crisis was short-lived; in the middle of 1943 a massive 'Big Inch' pipeline from the Gulf Coast to the north-east was completed a mere 350 days after construction work had begun, and by the end of the year the United States was putting tankers into service more quickly than the U-boats could knock them out. During 1944 the Battle of the Atlantic was eventually won, and by the end of the war every ton of allied oil-carrying capacity sunk by the Axis powers had been replaced twice over by American shipyards alone. The American national fleet numbered more than 800 tankers with a capacity equal to the entire pre-war world petroleum fleet.

The year 1944 was also a turning-point for the war in other theatres. In Europe Britain and the United States opened the second front with the invasion of France, while in the East the

* As at 30 June 1939.

Russians pushed steadily towards the German frontier. In Asia the American advance through the Pacific islands towards Japan made equally good progress, and with the capture of Saipan they were able to start bombing Tokyo itself.

For Germany and Japan intensive bombing made it virtually certain that within a short time they would run out of fuel. Neither country could import oil from overseas, and they had to rely increasingly on their synthetic plants. These were extremely efficient, and in 1944 more than 90 per cent of the Luftwaffe's petrol was derived from coal. But they were also very vulnerable to bombing, and between June 1944 and May 1945 allied air attacks reduced Germany's output of aviation fuel from 170,000 tons a month to virtually nothing.

The Japanese, with their customary ingenuity, managed to develop a process for producing aviation fuel out of pine-roots, but it was a poor substitute for the real thing. Early in 1945 their position was so desperate that they offered to hand over some of their few remaining cruisers to the Soviet Union if only it would supply them with oil. The offer was refused, but the fact that it was ever made showed how right Admiral Nimitz was when he re-arranged his priorities for winning the war. In December 1941, when he took over the American Pacific fleet, he said that victory would be a matter of 'beans, bullets, and oil'; in 1945 he decided that the correct order should have been: 'oil, bullets, and beans'.

CHAPTER 13

New horizons

With the arrival of peace the oil industry found itself on the threshold of a new golden age. It had played a vital role in winning the war, and it could be certain of a still more crucial position in the post-war international economy. The rest of the world had only to look at the developments that had taken place in the United States during the previous twenty years to gain some idea of what was to come; on the railways steam locomotives had given way to diesel; in the air the introduction of the DC3 had led to the creation of a comprehensive airline network; and in private houses at the time of Pearl Harbor new central-heating systems using oil and natural gas were being installed at the rate of more than a million a year. Europe was a heap of rubble, but once it had recovered, its people would want to enjoy the same comforts and amenities as the Americans.

New horizons were also opening up in the chemical industry. Until 1939 the oil companies had been so preoccupied with meeting the ever-rising demand for petrol, fuel, and lubricants that they did not have an incentive to take full advantage of the chemical potentialities of their raw material. The chemical industry for its part had been content to rely largely on its traditional feedstocks – coal, salt, wood, and animal and vegetable matter of one kind and another – and to purchase only small quantities of oil products for further treatment. The wartime shortages brought an end to this division, and threw the two industries together with sometimes spectacular results.

The most notable was rubber. When the Japanese overran Malaya, Britain and the United States would have been left without an alternative source of supply had it not been for the chemists' success in combining styrene with butadiene to form a synthetic substitute. They manufactured the butadiene from the butane which is to be found in natural gas and in the gases produced during the oil-refining process. Between them liquid petroleum refinery gases and natural gas also contain propane, ethane, and various other invaluable compounds, and it was realized that with the appropriate treatment there was almost no limit to the range of useful things that these could provide: toluene, the basic raw material for TNT, was one, and plastics another; there were also detergents to take the place of soap, fertilizers to increase the production of food, and synthetic fibres such as nylon. Much of the basic work was done before the war, but it is doubtful whether oil-based chemicals would have come into mass production so quickly if it had not been for the emergency.

Notwithstanding the pride with which these innovations are now described in the histories of the various companies concerned, the real source of wonderment is that they took so long to be made. As long ago as 1872, the famous Russian chemist Mendeleyev, after a visit to the Pennsylvania oilfields, told his government that: 'This material (oil) is far too precious to be burnt. When burning oil we burn money; it should be used as a chemical base material.'* There is, however, a big difference between a successful laboratory experiment and a large-scale commercial process, and in this instance it took a world crisis to bridge the gap.

Oil's† advantage as a chemical feedstock arises from the fact that it is principally composed of compounds of carbon and hydrogen, two of the most basic and versatile chemical elements. In the refining process these are broken down and thrown together into new syntheses. In the early plants they would either disappear into the atmosphere as gas or form liquids which had to be thrown away when no use could be found for them. The petrochemical industry was born when the waste products came to be regarded as possessing a value of their own, independently of the fuel oil, petrol, and lubricants with which they were being produced.

The most versatile are ethylene and propylene, some of the uses of which are shown in the following diagrams:

* *The Petroleum Handbook*, compiled and published by Shell.

† Natural gas, which has much the same chemical composition as oil, is dealt with in Chapter 22.

SOME USES OF ETHYLENE

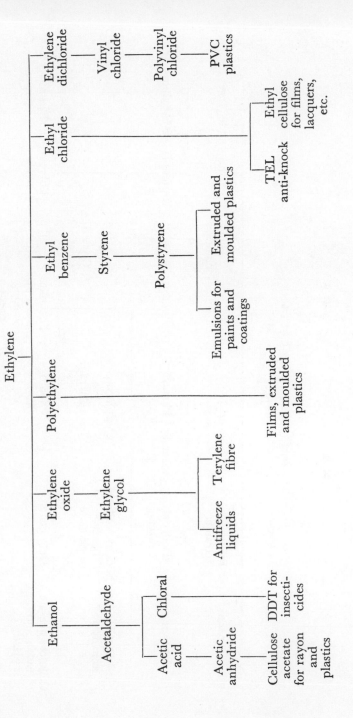

SOME USES OF PROPYLENE

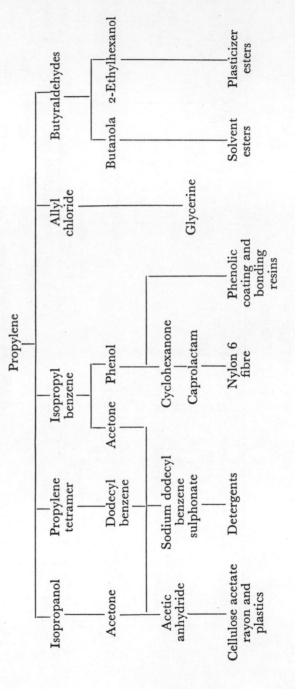

As soon as the guns stopped firing the American oil companies could get on with the job of exploiting their new opportunities. But in Europe this was impossible; nowhere, even in Britain, was there the raw material to build the necessary plants or the foreign exchange to import the oil from abroad. Only with American aid could industrial recovery begin, and the first generation of post-war refineries in Britain and on the Continent owe their existence to the Marshall Plan.

Yet for the oil companies the aid was a mixed blessing; on the one hand, it enabled Europe to buy the oil it needed and to build the refineries it would not otherwise have been able to afford; on the other, since these purchases were made at the expense of the American taxpayer, it made oil prices a matter of more general public concern than ever before. The days of private deals on a grouse moor were gone for ever; it had become necessary to justify one's arrangements before a government official or a Congressional inquiry. Under these circumstances the Gulf-plus price structure was bound to come in for criticism.

Indeed, the first attack had been made in 1943 by the British Government, when the Ministry of War Transport found that the price of bunker oils from the Persian Gulf for ships in the Indian Ocean and the Middle East contained a freight differential to bring it up to the price of supplies from the Gulf of Mexico. The matter was reported to the Auditor-General, who decided that 'the item should not be accepted', and the companies were told 'to furnish particulars for their actual f.o.b. returns for the period immediately preceding the outbreak of war'.* This they were unable to do, despite their usually admirable accounting systems, and the British Government never seems to have discovered the real costs of production in the Persian Gulf area. Nevertheless, the companies agreed to drop the phantom freight element from their prices and to establish a new basing point in the Middle East.

The first round could thus be said to have ended in a draw. The British Government was able to go ahead with its purchases in the knowledge that it was not being asked to pay for a journey

* 'Adjustments in Prices of Bunker Oil Supplies: British Auditor-General Outlines the Negotiations', *Petroleum Times*, 13 May 1944.

across the Atlantic that had never taken place; the companies for their part could congratulate themselves on the fact that at least the new basing-point price had been set at the same level as prices in the Gulf of Mexico. So the link between the Gulf price derived from the high American cost structure and the rest of the world with its much lower production costs remained, although in reality the contracts signed by the Allied armed forces often had lower prices.

In the immediate post-war years the connexion between the two base points led to some notable profits on the part of the companies. A typical example was a sale of crude oil and products from Saudi Arabia and Bahrain by California Standard and Texas to the American Navy for delivery to France under the lend-lease arrangements. Although the cost of production, including the Government's royalty, was only 40 cents a barrel in Saudi Arabia and 25 cents in Bahrain,* the company charged prices of $1.05 a barrel and upwards. When the deal was investigated by a Congressional committee in 1947–48 the Naval officers in charge of the negotiations were criticized for being 'far from diligent' in seeking cost records from the companies concerned.

Meanwhile the profits had gone on rising. The United States was in the grip of post-war inflation and passing through another of its periodic scares about its oil reserves running out. As soon as the wartime controls were removed in November 1946 prices in the Gulf of Mexico started rising. The Persian Gulf prices followed suit, and for the contract period December 1946–March 1947 the American Navy, buying on behalf of the United Nations Relief and Rehabilitation Agency (UNRRA), found itself paying Aramco $1.17 to $1.23 a barrel. More increases followed, and by March 1948 the price of Persian Gulf crude was standing at $2.22.

By this time Middle East production was running at 57m. tons a year, some $2\frac{1}{2}$ times the 1938 figure, and the situation began to change. After using Middle East oil to replace American supplies in many European markets the companies began importing it into the United States itself. Under these circumstances the Gulf price, and therefore all international prices, was bound to come down,

* Federal Trade Commission Report.

but the process was hastened by the launching of the Marshall Plan in April 1948, which led to the establishment of the European Co-operation Administration (ECA).

Under its administrator, Paul Hoffman, the ECA's job was to finance sales of oil and other vital products in Europe. Hoffman noticed that Middle East oil was undercutting American domestic oil as far inland as the mid-west, and concluded that the international companies must be selling to their American affiliates at lower prices than those being charged in transactions financed by the ECA. On 14 February 1949 he sent letters to the heads of Jersey Standard, Gulf, Socony Vacuum, and Caltex (a sales company jointly owned by California Standard and Texas), accusing them of differential pricing, and appointed a panel of experts to look into the matter. Already Middle East prices had dropped some way below their peak, but the fact that further cuts to $1.88 were made within months of Hoffman taking action seems to have been more than just a coincidence. Additional cuts were made soon after, and by the end of the year Persian Gulf oil was selling at $1.75, at which level the ECA continued to finance shipments until 1952, when the world oil shortage finally disappeared.

Even at this price, the profits were enormous, as production costs fell back after the completion of the large initial investments in well-head facilities, pipelines, loading terminals, and the rest. Various estimates have been put forward, and perhaps the most eloquent are those of the companies themselves, since these are the least likely to exaggerate. According to Standard Oil of California and Texas, the Aramco shareholders made net profits of 91 cents a barrel in 1948 and were still making 85 cents in 1950 when production was running at over half a million barrels a day.

No wonder the governments of the producing countries were becoming increasingly anxious to receive a larger share of the riches being extracted from under their soil.

CHAPTER 14

Venezuela leads the way to the 50/50 profit split

The demands for a greater share in the profits were led by Venezuela, and the precedents established there were almost immediately applied in the Middle East. In the years after the Second World War Venezuela was the largest and most sophisticated of the oil-exporting countries. It was also the one with the most experience of bargaining with the companies. It began to do so seriously in 1936 after the death of General Gomez, the dictator who originally allocated the oil concessions. In 1948 it introduced an income-tax law dividing the industry's profits equally between the State and the companies. The law grew out of Venezuelan conditions, and was designed to meet the needs of that particular country, but its essential provisions still form the basis for most of the world's more important concession agreements.

Gomez's principal service to his country was that he paid off its external debts and left it with one of the strongest currencies and largest monetary reserves in the world. But 'the opulent state' existed alongside 'an exhausted people'.* He entirely ignored the welfare of the great mass of the population; education was discouraged for fear that it would create political opponents, and little was done about health, despite the prevalence of tuberculosis, dysentery, malaria, and various other tropical diseases. The most conspicuous government services were the police and the armed forces.

After years of neglect and oppression the people welcomed the news of his death with riots and rejoicing. In the capital, Caracas, and in the oil town of Maracaibo the mobs raided government

* Cristobel Mendoza.

buildings and looted property belonging to the dictator's family. The foreigners who had financed his régime became a particular target; the Foreign Club was destroyed, while in the oilfields the women and children had to be evacuated, as the demonstrators tried to set fire to the wells and demolish the other installations.

For a while it seemed as if there would be a repetition of what had happened to Mexico after the overthrow of President Porfirio Diaz, with the country dissolving into semi-permanent civil war. But the care Gomez had lavished on his army proved not to have been in vain. The Minister of War, General Eleazer Lopez Contreras, soon quelled the disorders, and within a few weeks peace had been restored, with the General firmly ensconced in the Presidential Palace. The old order, it seemed, could continue, and General Isaias Medina Angarita summed up its feelings with the comment that 'if the army had not been united who knows what might have happened'. In the event Medina himself was one of the chief beneficiaries of this unity, for in 1941 he succeeded Lopez Contreras as president.

Nothing, however, could be quite the same again. When a dictator dies after many years in power it is never possible for his successor to go on as if nothing had happened. The people lose their sense of fatalism and inevitability. The very fact of the dictator's death shows that change is possible, and the Government is forced either to allow some relaxation or to risk another revolution. In Venezuela the military régime had no desire to make fundamental changes in the structure of society, but by demanding more money and other concessions from the oil companies it could simultaneously satisfy the popular desire for change while increasing the national income. Most educated Venezuelans recognized that their oil resources could not be developed without the international companies' help, but when nationalistic feelings are running high nobody likes to be dependent on foreigners. The more they are, the more they will resent it, and a Government which responds to this feeling can be sure of widespread popular support, even if there are far more urgent problems to be solved elsewhere.

The Lopez Contreras administration promptly moved against the oil companies on two fronts. It presented them with a list of demands for higher taxes and the ending of such privileges as their exemption from customs dues; at the same time it introduced a new Labour Law. To resist the financial demands the companies were forced to seek redress in the courts, thereby providing the clearest possible evidence that the industry and the Government were no longer the friends they had been in General Gomez's day, while the Labour Law demonstrated a new concern for the well-being of the people.

This concern was, however, more apparent than real. The oil companies were large employers, with about 25,000 workers, but these constituted less than 2 per cent of the working population. The vast majority were still employed on the land, and it was their needs which should have been given the first priority.

This is not to say that the oil workers could do without help. An investigation revealed that by North American standards there was a great deal wrong with their conditions. The houses in some of Shell's camps lacked bathrooms, Gulf had no hospital for its workers in the El Tigre region, while Jersey Standard's houses at Quiriquire were found to be too small. But in most parts of Venezuela nobody had ever seen a bathroom or heard of sanitation, and medical facilities were unknown. To the peasants even the worst of the camps seemed luxurious. They flocked to them from all over the country in the hope of finding work and being able to share in the benefits provided by the companies, until the oilfields were surrounded by shanty towns full of slum dwellers.

Rather than helping these unfortunates or the peasants on the large estates, the Government concentrated on making the companies do better what they were doing already. In addition to such mundane matters as wages, hours of work, and accident insurance, the 1936 Labour Law covered several areas which most wealthy governments normally regard as their own responsibility. The companies were obliged to build modern houses in permanent field camps, to provide a full range of medical services, and to educate the workers' children. The workers themselves were more interested in higher wages, and wanted to organize trade unions. But when a strike was called which cut production, thereby reducing the Government's income, Lopez Contreras took the companies' side,

Nothing could be permitted which might interrupt the flow of money into the treasury, and the unions were dissolved.

In economic affairs the Government was prepared to carry out more far-reaching reforms. Under Gomez oil's position in the economy was like that of a cuckoo in a nest full of sparrows. 'Between 1913 and 1935 Venezuela's exports, exclusive of petroleum, declined nearly 40 per cent in volume, and more than 50 per cent in value, whereas imports doubled in volume and rose three-fifths in value. . . . By 1935 oil represented more than 99 per cent of the volume and more than four-fifths of the value of Venezuela's exports.'* The prosperity of the country had become entirely dependent on oil, and thus on the level of industrial activity in the highly developed nations, an economic tyranny far harder to shake off than any political régime.

In the short run nothing could be done to alter this, but in the long run the oil revenues could be used to finance a comprehensive and balanced development programme. The name given to the policy was 'Sembrar el petroleo', since it involved sowing the fruits derived from oil in other enterprises. The main emphasis was sensibly given to agriculture, with special subsidies going to the producers of such traditional products as coffee, cacao, and hides, and a good deal of money was also spent on improving crops. The companies were willing to do a great deal to co-operate with re-form of this nature, and financed several projects on their own account.

Partly this was out of fear. In 1938 their long-drawn-out struggle with the Mexican Government ended in the expropria-tion of all foreign interests in that country. If Mexico, next door to the United States, could get away with nationalization it was clear that the danger could not be ignored in other countries. Some of Lopez Contreras' reforms were making life more difficult than under Gomez, but at least his government was one they could live with; far better therefore to co-operate than to risk having it overthrown and replaced by a more radical administration.

Nevertheless, fear was not the only emotion which prompted them. With their experience of European and American history the companies understood better than most Venezuelan officials

* *Petroleum in Venezuela*, Edwin Lieuwen.

the need to diversify the economy and to spread the wealth among the people. In both these areas they were among the most en-lightened employers, offering higher wages and pioneering pension schemes and shorter working hours in the knowledge that a contented work force leads through more efficient operating procedures to higher profits.* They were willing to apply the same thinking to Venezuela, where in any case they needed more trained nationals, and they consistently went further than their obligations under the Labour Law in the provision of scholarships and technical training so that locals could become drillers, engineers, geologists, and the like. The companies also realized the advantages of having the local population on their side and were willing to give help with many projects that seemingly had nothing to do with oil. Jersey Standard, for instance, gave materials for a church and an aqueduct to Quiriquire, while Shell presented the town of Maracaibo with a natural-gas system and supplied it with free gas.

There was only one thing the companies would not do to consolidate their position, and that was to pay higher taxes. There was still a world surplus of oil, with the threat of more to come from the newly discovered Saudi Arabian fields. In Venezuela itself wells were being 'shut in' because the market could not absorb their output, and there seemed to be no hope of prices rising in the foreseeable future. The companies therefore argued that they could not afford to pay more, and that it was unreason-able for the Government to ask.

To the Venezuelans the situation looked somewhat different. They could see only the enormous wealth of the companies, based, as it seemed, on their oil, and they wanted a larger share in the profits. Their attitude was eloquently expressed in 1938 during a debate in the Senate by Rivas Vasquez: 'We do not want to take away from the petroleum companies their equitable, reasonable, and just profits, but we want to defend our land and our Venezue-lan people by obtaining the just participation which corresponds to us from the exploitation of our national wealth.' Lopez Contreras shared this view, but the companies refused to give way.

* Standard's first pension plan in the United States started in 1903 and pre-dated social security by nearly thirty-five years.

The problem remained unresolved until the middle of the war, when the allies' desperate need for oil gave the Government, under its new leader General Isaias Medina Angarita, a much stronger bargaining position. A new agreement was signed in 1943 giving the Government an 80 per cent increase in revenue. The General was justly proud of his achievement, and saw it as the opportunity to answer the growing left-wing criticism of the military régime. In presenting the new royalty and tax arrangements to Congress he claimed that after years of exploitation by foreigners Venezuela would henceforth share equally in the oil industry's profits. The principle of equality was naturally greeted with enthusiasm, but instead of helping the Government, it provided fresh ammunition for the attack of the Opposition Acción Democratica which claimed that the companies were still receiving the greater share.

During 1944 and 1945 the Acción Democratica's expectations were proved correct, and in the autumn of 1945 the party acquired the power to implement its ideas after mounting a successful *coup* against the military government. This had nothing to do with oil. The country had become bored with military rule and anxious for more rapid social and economic advance. None the less, a change of régime was bound to mean a change of oil policy. The principle of an equal division of the industry's profits between the Government and the companies was a cardinal point in the Acción Democratica's programme, and the amendment of the 1943 agreement was high on its list of priorities.

The new Development Minister, Perez Alfonso, moved quickly to convince the companies of the desirability of coming to terms by imposing a special tax to bring their payments up to 50 per cent. Although this cost them $27m., they paid without any formal protest, and in 1946 the same thing happened again. With memories of the Mexican expropriation still fresh in their minds, the companies were worried about the possibility of even more extreme action, and they could not afford any interruption in Venezuelan supplies. In January 1948, after the first free elections in Venezuelan history had confirmed the Acción Democratica in office, the two sides started working out the details of an income-tax law that would guarantee the government its 50 per cent. The detailed discussions took several months, and it was not until 12 November that the law eventually came into effect. Within less than a fortnight there was another *coup* and the Acción Democratica was overthrown.

Once again the *coup* had nothing to do with oil. The Army was becoming restless under a civilian government, and disliked the growing power of the trade unions; it felt that the experiment in democracy had gone far enough, and it was time to return to the 'normalcy' of military rule.

The oil companies breathed a sign of relief as the Acción Democratica leaders slipped into exile. During their last months in office it had become apparent that their ambitions went far beyond the 50/50 profit split. They also wanted to secure an independent Venezuelan stake in the industry instead of leaving it entirely on the hands of foreigners. Already they had begun selling oil on their own account to buyers in the United States, Europe, and other Latin American countries, and plans were laid for establishing a national refinery. To the companies these seemed like the first steps along the road towards the establishment of a government monopoly. With the return of the soldiers the projects were put back in the filing cabinets to gather dust – but not for long. By the late 1950s similar ideas were being put forward in all the oil-producing countries as Venezuelans and the new generation of Arab oil experts, administrators, and politicians began to work together.

Until the income-tax law was actually on the Venezuelan statute book the companies always opposed the formal enunciation of the 50/50 principle. They felt that, as it took no account of their investments in developing the resources, it was a misleading formula, and they recognized that it could easily change into 60/40 or some other less favourable ratio, as indeed it has in many places.

However, once the principle had been adopted, they became its most enthusiastic adherents. They realized that, as it is easy to understand and seems so self-evidently fair to both sides, it could do much to create a better relationship with the Government. They also saw that such a revolutionary change could not be confined to one country without provoking resentment in the others, and that by extending the new deal to the Middle East they might, like the employer who offers higher wages before he is asked, gain a good deal of credit and more favourable terms than would otherwise be the case.

Throughout the Middle East in the early post-war years they were under pressure to increase their payments to the local governments. These knew how the value of the concessions had risen, and they were aware of the vast profits being made out of the oil produced in their countries. At the same time they resented the fact that through taxes the American, British, Dutch, and French Governments were taking a larger slice of the companies' profits than they themselves were receiving. The companies recognized the force of these arguments, and in the interests of maintaining good relations were anxious to accede to them. The dangers of not doing so were clearly revealed when two important newcomers broke into the area by offering the local rulers better terms than those incorporated in the major companies' concessions. In June 1948 Sheikh Ahmad of Kuwait granted a concession to Aminoil (a consortium of small American companies) for his share of the Neutral Zone* between Kuwait and Saudi Arabia, and in February 1949 the Saudi half of the zone was granted to J. Paul Getty's Pacific Western Company on conditions that were considered at the time to be much more favourable than those in the agreement with Aramco.

One possibility for the big companies would have been simply to increase their royalty payments to the host governments, but royalties could not be set off against their tax liabilities in their home countries. When, however, an American company paid income tax abroad it could reduce its tax payments at home by an equivalent sum. It was therefore agreed that although the companies should continue to make royalty payments on each ton of oil they produced (the rate was generally set at $12\frac{1}{2}$ per cent) the

---

* It was established by the Ojair agreement of 1924 when the Sheikh of Kuwait and Ibn Saud agreed that there should be a neutral zone along their frontiers in which the tribes could retain their grazing and watering rights and no forts would be built. It was expected that through the good offices of Britain firm frontiers would eventually be fixed and the zone would disappear. But this has never happened.

main increase should come in the form of taxes. As the Middle East countries did not have any income or corporate tax structure, the companies undertook to post an official export price known as the Posted Price. It then became a comparatively simple matter to subtract the cost of production and the royalty payment, and to divide the remaining profit equally between the two sides. The governments agreed. In 1950 Saudi Arabia became the first to implement the new system, and by 1952 all the other important producers in the area had followed suit with the exception of Iran.

The deals have proved much more favourable to the governments than could have been imagined in 1950. When the tax agreements were signed the posted prices were linked to the price of oil in the United States under the Gulf price structure. There was also a world shortage, and the price of refined products was correspondingly high. When, however, a surplus of oil came on to the international market in the middle and late 1950s and the price of refined products began to fall the governments refused to allow the posted prices to be reduced. The companies have had to introduce unofficial discounts while continuing to pay their taxes on the basis of the official posted prices. As a result 50/50 has in practice changed into something nearer 60/40.

CHAPTER 15

The Abadan dispute

In 1952 Iran was not merely the only country without a 50/50 profit-sharing agreement; it was also the only one where the Government had carried its disagreements with the concession-aire, in this case Anglo-Iranian, to the ultimate conclusion of nationalization. The subsequent trial of strength between the company backed by the British Government and the Iranian Government under the astonishing Dr Mohammed Mossadegh showed where the balance of power lay between the international oil companies and the governments of the producing countries.

It has also had a profound effect on the way relations between the two sides have since developed. On the one hand, the example of the damage caused by the Abadan dispute, as it is usually called, has done more than anything else to prevent disputes in other countries from getting out of control. On the other, it has led both companies and governments to follow policies designed to lessen their dependence on each other.

It was always more likely that a showdown would occur in Iran or Venezuela than anywhere else. In the Arab countries the industry was too much of a recent arrival, with the result that economic and political development was correspondingly less advanced.

But why should it have been Iran rather than Venezuela? Undoubtedly the country's history had much to do with it. Iran is among the oldest and proudest countries in the world, with a continuous history of civilization extending over twenty-five centuries. Like China, it has for long periods been a great power, and, as in China, its people felt deeply humiliated when they fell under Western domination in the late nineteenth century. The fact that this happened principally because of their own ineptitude increased rather than diminished their resentment, and during the first half of the twentieth century it found plenty to feed on as Anglo-Iranian grew to dominate the nation's economy.

The nature of the disputes between the Government and the company were much the same as those in Venezuela – taxes, royalties, customs privileges, and the rest – but they were always potentially more dangerous. In Venezuela the industry was made up of several companies of different nationality whose interests and relations with the Government were never entirely the same. Consequently, it never presented a single target for popular resentment. In Iran there was only Anglo-Iranian. The fact that the company was British also contributed to its long-term problems. Britain was the leading land power in Asia, through the Indian Empire, and after the collapse of the Turkish Empire it became the principal power in the Middle East. As such it was naturally feared and disliked by the local rulers as a threat to their independence; that was why King Ibn Saud of Saudi Arabia was so reluctant to give a concession to a British company. To Iranians Anglo-Iranian and the British Government were virtually indistinguishable, and economic and political disputes inevitably interacted on each other.

The worst was in fact political. It occurred in August 1941, when British and Russian troops invaded the country, without declaring war, to secure its communications for the supply by Britain of the Soviet armies in the Caucasus. Reza Shah abdicated to depart for exile in Mauritius* on a British warship, while his son, the present Shah, ascended the throne, an independent ruler in little more than name.

The end of the war and the withdrawal of foreign troops seemed to usher in a new age in which Iranians could at last become masters in their own house. The United States, with its well-known hostility to all forms of imperialism, replaced Britain as the most important country in the world; while all over Asia, in the Dutch East Indies, Indo-China, and above all India, local nationalists successfully challenged Europeans. Under these circumstances it needed no rabble-rousing politician or subterranean Communist plots to turn Iran against the company. As a symbol

* He was later moved to South Africa.

of British power, and of Iran's subjection to foreign influence, an attack on its position was inevitable under any government.

Anglo-Iranian was aware of the danger, and in 1948, nearly two years before the Americans introduced 50/50 to the Middle East in Saudi Arabia, the company opened discussions with the Iranian Government. The immediate cause of this initiative was the British Government's policy of dividend limitation. Under the terms of its concession (which had been amended several times since the original grant) the company not only paid the Government a royalty of 4s. a ton, like its counterparts in Iraq and Saudi Arabia, but also a sum equal to 20 per cent of any distribution to the ordinary shareholders in excess of £671,250 in any one year. Thus Iran's income was directly related to that of the ordinary shareholders, and when the Labour Government in Britain held down their dividends the Iranians suffered as well. The company offered to make this good through a straight cash payment, but the Iranians felt that it ought to form part of a more far-reaching change in the whole agreement.

During the talks a 50/50 deal on Venezuelan lines was considered. But when the Iranians, reviving a traditional demand, insisted that it should apply to all the company's profits, whether derived inside or outside Iran, Anglo-Iranian refused. Instead a 'Supplemental Agreement' to the existing concession was negotiated. When it was signed on 17 July 1949 by Mr (later Sir Neville) Gass on behalf of the company and by the Iranian Finance Minister Mr Golshayan its terms were hailed as the most generous in the Middle East. The royalty payments were to be increased to 6s. a ton, the Iranian Government's share of the profits was guaranteed never to fall below £4m. a year, and the company promised an immediate payment of £5m. as compensation for the effects of dividend limitation.

The Iranian National Assembly, however, was dissatisfied. There were only ten days left for debate before it had to be dissolved for new elections, but this was enough for a group of opposition deputies led by Mossadegh to discredit the new arrangements. His arguments were devastatingly simple: the company was plundering Iran of its raw materials and must be made to pay for them so that the wealth of the country could be spent on its own people instead of by foreigners.

Mossadegh was already over seventy with a long and chequered career behind him, during which he had at various times been Minister of Finance, Justice, and Foreign Affairs, and imprisoned by Reza Shah for plotting against the Government. As a landowner related to the former Qajar ruling family, he never displayed much interest in social affairs; but he had always been a passionate nationalist, with a vague, romantic conservative view of how Iran should be governed in order to retain its independence; this at one time had led him to oppose the building of railways lest they bring the country into closer contact with Britain and Russia; Anglo-Iranian, as a foreign company dominating the economy and introducing alien customs, epitomized everything he abhorred. Under the prevailing circumstances Mossadegh's nationalism would by itself have been enough to give him a strong voice in Iranian politics; but the situation was ripe for more than that. Iran in this period bore a close resemblance to the French Fourth Republic, with a large Communist party called the Tudeh and political leaders playing musical chairs with the premiership. The Shah had not yet emerged as the strong man he is today, and the field was wide open for a politician who could build up a genuine power base. Mossadegh saw the opportunity, and by skilfully linking the oil issue with the causes of nationalism and social reform was able to make himself irresistible.

In February 1950, when the new National Assembly opened, his strength was deceptively weak; his new National Front party, founded a few months previously, had returned a mere eight members. But through his chairmanship of the National Assembly's Oil Committee, the widespread appeal of his arguments, and his growing control over the Tehran mob he quickly became a key figure. 'How could anyone be against Mossadegh?' the Shah has since asked,* 'he would enrich everybody, he would fight the foreigner, he would secure our rights. No wonder students, intellectuals, people from all walks of life flocked to his banner.' If they did not they risked being beaten up or branded as a traitor in the increasingly hysterical atmosphere of Tehran.

* *Mission for my Country*, Reza Shah Pahlavi.

The Government under the right-wing General Ali Razmara wanted to settle the dispute with the company peacefully through revisions to the Supplemental Agreement so that it could press ahead with a programme of constitutional reform. This was altogether too cautious for Mossadegh, and by the end of the year he had forced it to renounce the agreement entirely. In January 1951 his position was further strengthened by the announcement of Aramco's 50/50 agreement in Saudi Arabia. In an attempt to keep oil out of politics Razmara had asked the company not to publicize the advantages of the Supplemental Agreement, and most Iranians had no idea that it would have yielded as much as a 50/50 deal. They knew only what they had heard from its opponents, and the settlement in Saudi Arabia confirmed all their suspicions that Britain was not treating them fairly.

Anglo-Iranian immediately suggested talks leading to a 50/50 settlement in Iran plus a down-payment of £5m. and monthly instalments of £2m. as advances against future royalties. But it was too late. On 19 February 1951 Mossadegh formally suggested outright nationalization. Fearful of Mossadegh's growing power, and aware of Iran's inability to run the industry without foreign help, Razmara tried to play for time by referring the proposal to a panel of experts, which rejected it as impracticable and possibly even illegal.

The result was to throw Tehran into a state of chaos, as the National Front launched a series of demonstrations against the Government. On 7 March Razmara was assassinated on his way to the Mosque by a religious fanatic, and thereafter events moved rapidly. On the following day the frightened Oil Committee backed Mossadegh's plan for nationalization, and on 15 March, a simple nine-point Bill nationalizing the industry was passed through the National Assembly. The Shah, fully sharing Razmara's fears, appointed the elder statesman Hussein Ala as Prime Minister in a last desperate attempt to find a compromise solution. By this time, however, the situation was completely beyond his control, with strikes and civil disorders spreading to the oilfields and the refinery at Abadan. By 1 May he had been forced to give his Royal Assent to the Nationalization Bill and to appoint Mossadegh as Prime Minister.

Most Iranians had little idea of the implications of the enterprise on which they had embarked. They shared with Mossadegh the belief that the British economy would collapse without Iranian oil so that Britain would be forced to go on buying, and they thought that the American companies would be only too delighted to take advantage of Anglo-Iranian's discomfiture. Once a national oil company had been established and Anglo-Iranian given whatever compensation the Government in its wisdom might decide, they believed that everything would go on as before, except that there would be no foreigners to pre-empt the profits.

Neither Mossadegh nor his advisers had any conception of the problems involved in running a major company, nor the faintest idea of how to sell their oil on international markets without a tanker fleet or distribution system of their own. Because they thought that Iranian oil was even more vital to their customers than it was to Iran, they supposed that they could do as they pleased, rejecting all compromise proposals. It was this over-confidence which was to prove their undoing. They failed to realize that Britain could draw on alternative reserves in the Arab countries, Venezuela, and the United States whereas Iran had no alternative source of income.

Nevertheless, the British were quite prepared to compromise, and both the company and the Government put forward several proposals accepting the principle of nationalization. The first came from the company in June, when the deputy chairman Basil Jackson visited Tehran. He opened his discussions with an offer of £10m. down and £3m. a month until a settlement was reached as an advance against royalties, and then outlined his proposal. In essence, it was that Anglo-Iranian would establish a new company, with Iranians on the board, to operate the country's industry on behalf of the newly established state-owned National Iranian Oil Company. The Iranian delegates rejected the scheme out of hand after half an hour's consideration, and proposed that the National Assembly and Senate should be the sole judges of the compensation to be paid for the take-over. In August, after President Truman's special representative, Averell Harriman, had had talks with Dr Mossadegh, a British Government mission under Mr Richard Stokes, the Lord Privy Seal, went out to Tehran with a basically similar plan. He suggested that a new company should be established to run the industry on behalf of the National Iranian Oil Company and that the national company should sell oil for export to Anglo-Iranian at a price designed to give an equal share in the profits to both sides. Again the Iranians refused to consider the proposal.

Nothing further could be done except to appeal to the International Court at The Hague, whose jurisdiction the Iranians anyway refused to accept, and when Dr Mossadegh insisted that the British staff in Iran should either work for his company or leave the country they were withdrawn. The last of them left on 3 October, ironically enough in a warship called H.M.S. *Mauritius*, and operations in the oilfields and at Abadan were brought virtually to a halt. With a certain amount of foreign help they could have been kept going. But that was not the point. To their horror, the Iranians discovered that they had been cut off from their markets. Nobody any longer wanted their oil.

The days were past when Britain could send a gunboat to defend its commercial interests. But by his refusal to agree to any reasonable compromise or to suggest any satisfactory scheme for compensation, Mossadegh exposed his country to an even worse threat: an embargo which no other country was prepared to break. On the day after the last Briton left Abadan the Foreign Office in London announced that it would do everything possible to prevent the sale of Iranian oil; when Italian and Japanese concerns made small purchases from the National Iranian Company, Anglo-Iranian brought lawsuits against them, and these were enough to prevent further sales.

Iran was literally subjected to a slow strangulation. In the two years before the expropriation it exported 54m. tons of oil; in the next two, sales amounted to 132,000 tons, yielding less revenue than a single day's royalties under the old company. At the same time the Government had to take over numerous responsibilities formerly carried by the company. Some of these were directly connected with oil, such as the 70,000 Iranian oil workers being paid £20m. a year; others, like the 1,250 miles of road and forty major bridges built and maintained by the company at its own expense, had little to do with the industry.

Britain and the company by contrast recovered from the initial shock surprisingly quickly. When the blow fell Iran accounted for about three-quarters of Anglo-Iranian's crude oil production and refining capacity, but the company also had a half-share in the Kuwait Oil Company, and 23⅓ per cent in the Iran Petroleum Co., which had extended its interest beyond Iraq to the Sheikhdom of Qatar. These easily filled the gap left by Iran. In Kuwait alone output rose from 17m. tons in 1950 to 42m. tons in 1953, while in Iraq over the same period the increase was from 8·1m. to 27m. tons. In Britain and the other European countries normally supplied by Iran valuable dollars had to be spent buying crude and products from the United States and Venezuela. But by the end of 1952 it was already apparent that the combined effects of the increases in the Middle East and the construction of new refineries in Europe would shortly enable the area to do without American help.

Nevertheless, Mossadegh still refused to compromise. Early in 1952 the World Bank offered to set up a temporary management for the Iranian oil industry so that negotiations for a final settlement could take place. The initiative collapsed over the Iranians' insistence that the bank should act as their agent instead of as a neutral intermediary and their refusal to accept British technicians. Joint Anglo-American proposals put forward in August 1952 and February 1953 met with equal lack of success through failure to agree on the principles for compensating Anglo-Iranian. If agreement had been reached on this point in February 1953 the American Government was prepared to give Iran immediate cash grants to be repaid in oil in order to get the industry going again. Britain had meanwhile moved a long way from the position of 1951. No longer did it demand that Anglo-Iranian should have a monopoly of the distribution of Iranian oil on world markets. The company was prepared to see this task undertaken by an international consortium of which it would only be one of a large number of shareholders.

By this time Mossadegh had succeeded in dissipating most of his support within Iran. The economy was in ruins, with rising unemployment and galloping inflation. The political situation was even worse, with the Supreme Court dissolved, the Senate abolished, and the National Assembly suspended. Government had come to mean Mossadegh and the mob supported by the Tudeh Party. The country was moving towards a dictatorship, which seemed likely to be merely a prelude to Communism. In July the Shah tried to dismiss Mossadegh from office, and was himself forced to flee to Rome. *En route* he had to come down in Baghdad, and so shaky did the monarchy's position look that the Iranian ambassador there tried to have him arrested. It was an unwise move. Mossadegh had gone too far. Within a few days a military *coup d'état* led by General Zahedi had swept him from office, and the Shah returned in triumph.*

The restoration of the oil industry could not be managed as quickly, although discussions began almost immediately on the understanding that a consortium of international companies should be formed to take the place of Anglo-Iranian and work on behalf of the National Iranian Company. Not until 29 October 1954 did the new agreement come into force, but within a month oil was flowing through Abadan at the rate of 6m. tons a year.

On paper Iran looked much better off than in 1951. The principle of nationalization was recognized, and the National Iranian Company retained its ownership of the former Anglo-Iranian concession. The new consortium produced the oil on its behalf, but the oil remained the property of the National Iranian Company, which then sold it to the individual members of the consortium and shared the profits equally with them. The consortium included all eight of the world's leading international companies – Anglo-Iranian under its new name of the British Petroleum Company,† Shell, Jersey Standard, Socony Mobil,‡ California Standard, Gulf, Texas, and the Compagnie Française des Pétroles.§ There could therefore be no question of Iran any

* It is frequently alleged that the American and British secret services financed the rising, and it is perhaps significant that in his memoirs the Shah leaves the question open.

† Always called BP.

‡ The former Socony Vacuum.

§ The American companies were shortly afterwards obliged by their government to give a 5 per cent holding to a number of independents, formed for the purpose into the Iricon Agency. This was to avoid the impression that the consortium was a cartel.

longer being unduly dependent on Britain. Yet for all its complexity the agreement did not provide Iran with a larger share of the profits than the 50/50 deals negotiated by Iraq, Saudi Arabia, and Kuwait a few years earlier. In the meanwhile Iran had lost valuable markets to those countries, and its position as the leading producer in the Middle East had been taken over by Kuwait. Moreover, there was no compensation for the loss of earnings during the period of the shut-down.

The position of Anglo-Iranian was quite different, with the Iranian Government and the other members of the consortium agreeing to pay it compensation. From the Government the company received a promise of £25m. to be spread over ten annual payments beginning in 1956 to cover the seizure of its assets, and from its partners £32·4m. plus an overriding royalty of 10 cents on every barrel of oil exported until a total of $510m. had been reached. 'It is an awful blow to wake up and find that you have lost most of your oil,'* said Lord Strathalmond, the company's chairman, soon after the nationalization decree in 1951. In the long run, however, it can be seen to have done the organization much good; Mossadegh forced it to stop putting all its eggs in one basket, to develop new oilfields in the Middle East (particularly Kuwait), and to look for new reserves outside the area in North Africa, Canada, and elsewhere, so that it could never again be held to ransom by a single government.

The rest of the industry drew the same conclusions from the Abadan dispute. Although by the early 1950s it was clear that the Middle East contained the largest oilfields in the world, with the lowest production costs, no major company wanted to rely on the area alone for its supplies; nor did the Government of any oil-importing country. As a result, the industry started spending millions of pounds looking for new fields in other parts of the world, a search that was intensified after the Suez crisis of 1956 showed how easily the supply of Middle East oil could be interrupted.

The Governments of the producing countries also learnt valuable lessons from Abadan. The Shah went to the heart of the matter when he explained that Mossadegh's 'big miscalculation

* Longhurst, op. cit.

lay in his stubborn insistence that he knew how to market our oil with no help from foreigners. Yet at that time we possessed not a single tanker, nor did we have even the beginnings of an international marketing organization.'*

In one way or another all the Governments have since adopted policies designed to remedy these defects. That is why they have established national oil companies to give themselves experience of how the industry works right through from the production of crude oil to the sale of the finished products. At the same time, just as the international majors have been trying to diversify their sources of supply, the Governments have been looking for new concessionaires to provide alternative sources of revenue. Accordingly, they have welcomed opportunities to grant concessions to the American independents and the state-owned organizations from Europe which began to appear on the international oil stage as the 1950s wore on.

* *Mission for my Country.*

CHAPTER 16

The arrival of the newcomers

'If one is to be anybody in the world oil business, one must have a footing in the Middle East.'

J. Paul Getty

One of the strengths of the free-market system is that it is self-correcting, and this is as true of the international oil industry as any other. Whenever there is a sellers' market and prices are high new companies are attracted into the business, leading to the discovery of new fields, more competition, and lower prices. This happened in Europe during the 1880s, when Russian oil prevented the Americans from gaining a near monopoly, and again after the First World War, when the Americans found new fields at a time when it looked as if the British had gained control of most of the world's remaining reserves. The pattern was repeated during the late 1940s and early 1950s.

On this occasion the newcomers fell into two categories: independent American companies such as Continental, Marathon, Phillips, and Signal, and organizations from the consumer countries that were either owned directly by the state, like the Italian ENI* or the French BRP,† or backed by their parent government, such as the Japanese-owned Arabian Oil Company. Whatever their origin, these companies had fundamentally similar motives: on the one hand, they or their backers were afraid of the enormous power conferred on the majors by their almost unlimited reserves of cheap Middle East and Venezuelan oil, while, on the other, they wanted to earn similar profits for

* Ente Nazionale Idrocarburi.

† Bureau de Recherches de Pétroles. On 1 January 1966 it merged with another state-owned concern, Régie Antonome de Pétrole, to form the Entreprise de Recherches et d'Activités Pétroliènes, always known as ERAP.

themselves. It is true that the American companies, as commercial concerns, were the more interested in profits, and the consumer country organizations, having access to government assistance, placed a higher priority on breaking the stranglehold of the majors on their markets; but this was purely a matter of emphasis.

In terms both of their numbers and the scale of their operations the American independents were the more important group. For many years before the war some of them had operated outside their own country; Standard of Indiana, for instance, was one of the biggest companies in Venezuela before selling out to Jersey during the depression,* and in 1945 there were thirteen of these companies with foreign exploration or production interests, mostly in Canada and Latin America. As a group, however, they counted for practically nothing beside the big seven. By 1958 the situation was completely different and the number had risen to over 200; by the early 1960s some of them, most notably Continental and Marathon, which have big reserves in Libya, were well on the way to becoming majors themselves.

Different companies acted for different reasons and at different times, but since all of them were in competition with the international majors in the United States and responding to the actions of the majors, the environment in which the boards of directors took their decisions was basically the same in each case. The first to make an impact on the international scene were Getty and the Aminoil Group, both of which in 1948 acquired concessions in the Neutral Zone between Saudi Arabia and Kuwait, but in the long run the most important has turned out to be Continental.

In 1948, when the company's board under the newly appointed president, Leonard F. McCollum (always known throughout the industry as Mr Mc), took the decision to expand its exploratory effort beyond the United States it looked small beside Jersey Standard, Shell, or Gulf; but in the oil industry a relatively small company is frequently very large by any other standard; Continental was among the fifty largest corporations in the United States with total assets of over $250m. and an after-tax income of over $30m. a year. Its crude-oil production and sales of

* See Chapter 10.

refined products were more in balance than those of any other oil company in the United States, and with the economy booming and oil demand rising rapidly it could look forward to fat profits for as far ahead as the eye could see.

For all this strength, the board could see important weaknesses in the company's structure; in 1948 they were no more than hair-line cracks in a piece of metal, but in the long run they looked as if they might develop to the point where they could prove fatal to its whole competitive position.

It was all a matter of looking at the direction in which the lines on various graphs were moving. To begin with, there was the rising cost of finding and developing new oil reserves in the United States; in 1946 the average cost per barrel had been 84 cents, by 1948 it was already over $1, and there was every reason to suppose that it would continue to climb. At the same time the international companies were importing increasing quantities of low-cost foreign oil into the United States, which they were selling below the price of domestically produced oil, as Mr Paul Hoffman of the ECA pointed out.* In 1948 the United States found itself in the unfamiliar situation of being a net importer of oil, and it was obvious that as production costs in the Middle East fell back after the initial capital investment had been completed the volume of imports would rise even more rapidly unless the Government took action. The company therefore felt that if only to protect its existing markets it, too, had to find a source of low-cost foreign crude. 'There I sat,' McCollum explains, 'I had no hope of stopping the flow of foreign oil. I didn't want to rely on politics. So why should I worry if I bought in ten barrels from abroad and lost ten barrels at home? I decided that if we didn't go abroad we'd end up as another insular company.'†

There were carrots as well as sticks to be taken into account. Because the Middle East and Venezuelan prices were linked to those of the United States through the Gulf-price structure the international companies were earning enormous profits on their foreign operations. There was also the promise of overseas markets to be considered. Once the Marshall Plan had taken effect, there was every reason to suppose that the rate of growth for oil in

* See Chapter 10.   † *Management Today*, April 1967.

7. Crew working on the drilling bit, Lake Maracaibo

8. North Sea drilling platform

Europe would start to rise more rapidly than in the United States. All the American independents who went abroad intended to bring most of the oil they discovered back to their domestic market, but they also hoped to take advantage of any opportunities for making money abroad that might arise.

However, the arguments were not all on one side. The cost of finding oil in the United States might be rising, but at least it was known to exist; foreign countries might look promising and then swallow up large sums of money without yielding anything in return. In addition, many of them were politically unstable and suspicious of foreign capital.

Not surprisingly, the independents tended to concentrate at first on Canada, which looked very promising after the discovery of a big field at Leduc in Alberta in 1947, and Venezuela, and to approach other areas with caution. This led to the formation of joint-venture companies as a means of spreading the risk; Continental went in with Marathon and Amerada. In 1948 they began exploring in Venezuela, and then, having got their feet wet, extended their operations to North Africa, where they eventually discovered vast reserves in Libya in 1959.

Many other companies followed a similar course, though not always with so much success, while a few lucky ones had the Government to do the hard work for them. Although American oilmen always claim that they want nothing more from their Government than to be left alone, in reality they rely on it heavily for taxation privileges and assistance of various kinds. One of the clearest examples of this occurred in 1954 after the formation of the international consortium in Iran. The American authorities were afraid that the presence of so many international majors in a single enterprise would look like a cartel. So they made the five American members sell 1 per cent each of their holdings to nine independents formed for the purpose into the Iricon group.* The oil was already found, and like Gulbenkian in the Iraq Petroleum

* The members were Aminoil, Richfield, Ohio Standard, Getty, Signal, Atlantic, Hancock, Tidewater, and San Jacinto. Getty and Tidewater are both members of Mr J. Paul Getty's group. Atlantic and Richfield have since merged, while Continental has taken over San Jacinto, and Signal has taken over Hancock.

group, they were not expected to put any effort into running the enterprise. After making the initial investment they merely had to enjoy their profits, and their experience did much to stimulate interest in overseas areas among the other independents.

The organizations from the consuming countries were rather slower off the mark than the Americans for the obvious reason that after the war their Governments were desperately short of capital. The Germans and the Japanese were, of course, in a hopeless position to do anything, and at first the running was taken up by the French and the Italians.

The French were in the stronger position of the two, since in the Compagnie Française they already had a major oil company of their own. But the Compagnie Française needed all its resources to develop its share of the Iraq Petroleum operation and to rebuild its war-shattered facilities in Europe. It had nothing to spare for exploration ventures elsewhere. The Government, for its part, was anxious to break away from its dependence on the Middle East by trying to find oil in French territory, especially the Sahara. Its reasoning was the same as Churchill's when he made the British Government part-owners of Anglo-Persian in 1914; oil was now so vital to the economy that France had to have a supply under its own political and commercial control. In the Iraq Petroleum Company the Compagnie Française could do nothing without the consent of its partners, and in practically every dispute it found itself in a minority of two with Gulbenkian against the Anglo-Saxons. Moreover, France no longer had any direct political influence left in the Middle East.

In 1945 the new Directeur des Carburants, Pierre Guillaumat, established the wholly-government-owned BRP to invest money in companies prepared to prospect for oil anywhere in the French Empire. The majors were doubtful of its chances, but as in Texas and the Persian Gulf, the visionaries outside the established oil industry who said that oil could be found in an untried area were proved right. In 1955, a year after the start of the Algerian revolution, the first discovery was made in the Sahara at Edjeleh, and in 1956 another field was found at Hassi Messaoud, and a vast natural-gas field at Hassi r'Mel. Despite the war, which made it extremely difficult to build and defend the pipelines to the coast, the Saharan oil and gas reserves were developed extraordinarily quickly, and when Algeria achieved its independence in 1962 the new Government took over a thriving industry with production running at 20.7m. tons a year.

While the French were able to concentrate on their empire without disrupting the rest of the international oil scene, the Italians had to take their chances where they could. During the Abadan dispute Enrico Mattei, the head of the state oil and natural-gas concern, ENI, had co-operated with the international companies by refusing to buy oil from Iran, and when the consortium was established he had expected to be rewarded with an invitation to participate. None came, however, and when he asked why he was told that membership was being confined to those companies which already held concessions in the Middle East. Within a few months the formation was announced of the Iricon group of American independents, none of whose members held any such concessions. To Mattei this seemed like a deliberate snub. He never forgave the major companies for the rebuff, and he determined that through him Italy should secure a place at the top table of international oil affairs in its own right. His struggle to get there* made him a world-famous celebrity as a sort of latter-day David challenging the largest of all contemporary industrial Goliaths before he was killed in an aeroplane crash on 27 October 1962.

The heads of the international oil companies looked on Mattei in much the same way as a group of Roman senators might have regarded a barbarian chieftain; he was a throw-back to their own barely remembered past. By the 1950s these companies had long since forgotten the heroic age when they were run by ruthless, single-minded individuals such as John D. Rockefeller and Henri Deterding; men whose power and fortune rested on their ability to exploit new ideas and to overturn the established order. They were controlled by technocrats and managers who had worked their way up through recognized promotion structures and were accustomed to conducting their affairs in the ordered calm of air-conditioned board-rooms with departments of public-relations men to protect them from the curiosity of the outside world.

* The best account is contained in *Mattei: Oil and Power Politics*, by P. H. Frankel, who knew Mattei and advised ENI.

Mattei had more in common with the companies' founders than with their successors. He was an outsider, a former resistance leader, who had come into the oil industry by accident after the war when the Government made him Northern Commissioner of the Azienda Generale Italiana Petroli (AGIP), the most important of the enterprises set up by the Fascists to participate in the oil and gas industries. His reputation was made by AGIP's discovery of natural gas in the Po Valley in 1949, and in 1953 he persuaded the Government to bring together all the state holdings in oil and gas in the Ente Nazionali Idrocarburi (ENI) under his presidency. As a former Christian Democrat Member of Parliament, Mattei was as much at home in politics as in business, and as the owner of the daily paper *Il Giorno* he had direct access to the public, which helped him to present his fight with the oil companies as a crusade on behalf of Italy. Like Deterding, he was a sort of freelance industrial buccaneer who just happened to be in the oil business: he could just as easily have been a Cecil Rhodes opening up a continent, or a David Lloyd George destroying the Liberal party to further his personal ambition, had he been born into different circumstances.

Also like Deterding, Mattei was a revolutionary who devised a new method in order to overcome his fundamental weakness. Deterding's handicap had been his lack of indigenous supplies with which to tackle the Americans; so he developed a system of always acquiring alternative sources of oil as near as possible to his markets. Mattei's problem was that Italy was a latecomer to the international oil scene, and one without international influence. In 1914, when Churchill launched the Anglo-Persian, Persia was virtually a British dependency, and between the wars, when the American and British companies had carved up the Middle East, the concessions had been allocated as part of an inter-government settlement. The Compagnie Française also acquired its stake in Iraq through the good offices of the French Government, while BRP and its associates in the Sahara were exploring in French territory. ENI started with no such advantages, and to make matters more difficult the majors were more powerful than ever before. As Iran had discovered to its cost, only the majors could sell large quantities of oil throughout the world, and ENI did not have the financial resources to bid against them for concessions by offering more generous financial terms.

Mattei saw that the way round his dilemma was to put his offers onto a completely different footing from those of his rivals by appealing to the nationalistic sentiments of the producer countries. 'The people of Islam are weary of being exploited by foreigners,' he announced. 'The big oil companies must offer them more for their oil than they are getting. I not only intend to give them a more generous share of the profits but to make them my partners in the business of finding and exploiting petroleum resources.'*

In August 1957 he put these ideas into practice when he signed a deal with the Iranian Government. It was the first to break away from the established 50/50 pattern, and the first to make the host government a genuine partner rather than just a tax gatherer. Its basic terms were very simple: ENI formed a joint company with the National Iranian Oil Company, called SIRIP,† and agreed to share both the management decisions and the future profits equally with its partner. In addition, ENI agreed to pay 50 per cent of its profit in tax. As the National Iranian company was a state-owned concern, this meant that altogether the Government would receive 75 per cent of SIRIP's total profits. The major oil companies were quick to point out that these terms were by no means as generous as they looked. To begin with, a major company would have paid a substantial down payment, probably as much as $40m., for such a promising concession, whereas ENI paid nothing. Secondly, in a conventional concession agreement the foreign company pays all the exploration and development costs, whereas ENI expected the National Iranian Company to repay half of these if oil and natural gas were found, though not if the venture ended in failure.

Nevertheless, the majors regarded the SIRIP deal as a dangerous innovation, and the United States ambassador in Tehran called on the Shah 'to discourage us from entering into this new kind of agreement'.‡ They had always valued their commercial

---

* *The Politics of Oil*, Robert Engler. Macmillan (New York).

† Société Irano-Italienne des Pétroles.

‡ *Mission for My Country*.

freedom very highly, and it was bound to be reduced in any partnership with a government-owned concern. They also saw that once established the precedent would quickly spread, and then the terms would again have to be improved in the Government's favour. The Shah entirely agreed with this analysis. He had little faith in ENI's ability to find a large field, and his judgement on this score was soon proved correct. His aim was simply to establish a precedent which the larger companies would have to follow, and he intended to play them off against the newcomers in the best interests of Iran.

The plan worked perfectly. The prospects for making a large discovery in the Middle East were far too promising for any remotely reasonable alteration in the concession agreements to deter a new company from entering the area. In December 1957 the Japanese-owned Arabian Oil Company acquired a concession off-shore from the neutral zone between Kuwait and Saudi Arabia in which the Governments of those two countries were given a stake, and a few months later the Iranian Government signed another partnership deal, this time with Standard Oil of Indiana. Both these deals contained further important innovations in favour of the host government. The Arabian Oil Company agreed to share the profits on refining and marketing outside the production area with the Saudi and Kuwaiti Governments, while Indiana Standard paid the Iranians a cash bonus of $25m. on the signature of its contract. It also promised to spend $82m. on exploration over a period of twelve years, and to share any saving with the National Iranian Company if it was successful before the full amount had been spent.

Not even the major companies could stand out against the trend indefinitely. In their annual reports and chairmen's statements they fulminated against the injustice and unwisdom of the changes, holding up the 50/50 principle as the model for fair dealing. None the less, they could not allow all the most promising new concession areas in the Middle East to pass into the hands of outsiders. British Petroleum, Gulf, Texaco, and other companies with enough oil to meet their requirements could afford to stand firm, but others could not. Since the war Shell had been notably unsuccessful at finding enough oil to feed its enormous distribution system, and in the late 1950s its need for new supplies was far more important than any esoteric commercial principle. In 1960 it became the first major company to accept a host government as a partner when it took out a concession for Kuwait's off-shore areas. It promised that following a commercial discovery the Ruler would have the option to acquire 20 per cent of the shares in the venture, entitling him to 20 per cent of the production. Shell agreed to look after the sale of this oil on his behalf, but it was clearly understood that in due course it would be used to launch the newly formed Kuwait National Petroleum Company into world markets. Four years later in 1964 Shell went a step further to match the terms introduced by Mattei when it established a joint company with the National Iranian Oil Company to develop the Iranian off-shore areas. Thus the revolution of 1957 achieved the accolade of respectability.

CHAPTER 17

The formation of OPEC

Both the major companies and the newcomers were successful at finding oil beyond their wildest dreams. In the late 1940s when the newcomers began to appear on the international scene, there were still lingering fears that one of the problems of the post-war world would be a shortage of oil, and reserves were sufficient to last for less than fifteen years at the prevailing rate of production. By 1960 there was enough to last for over forty years, despite the vast increase in consumption, and the industry's main problem was to find ways of preventing the flood of new discoveries from washing away their profits in a wreckage of price wars.

The Middle East concessions acquired before the war by the major companies turned out to contain far more oil than anybody had expected, and large new fields were also found in many of the post-war concessions. Some of these were located in the traditional oil-producing areas of the Persian Gulf and Venezuela, and others were discovered in entirely new places. In Africa the French found massive reserves in Algeria, next door in Libya a whole host of companies were successful, and on the other side of the Sahara Nigeria became a major exporter. It is perhaps not surprising that oil should have been found in Africa, since the Continent had never before been subjected to a thorough exploration. But quite large fields were also uncovered in areas which had been closely examined several times in the past, such as Canada and Australia.

When prices began to collapse under the weight of oil coming on to the market it was assumed in some quarters that the major companies would resume their efforts to form an international cartel. There are indications that Enrico Mattei took this view, and part of the explanation for his aggressiveness may have been a desire to establish a strong enough position to qualify for membership. The idea seems to have been put forward in the middle-management reaches of some of the companies, but there was never any chance of its acceptance at the top level. In August 1952 the United States Federal Trade Commission published a detailed and hostile report * on the industry's activities during the 1930s and '40s, and the American companies knew that any move on their part to restrict competition would lead to a disastrous clash with the Government.

Moreover, the climate of opinion within the industry was quite different from that of the 1930s. In the aftermath of the depression all businessmen were naturally on the defensive and afraid of the effects of too much competition; in the 1950s the world economy was booming, and there was a general feeling of confidence in nearly all industrial circles. Oilmen could feel particularly optimistic as the demand for their product bounded ahead. In Western Europe alone oil's sales increased from 97m. tons in 1954 to 201m. tons in 1960, and over the same period its share of the total energy market rose from less than 20 per cent to about 35 per cent.

Under these circumstances the heads of the major companies were more concerned to increase the turnover of their individual companies than to restrict competition, and the fact that production was outrunning demand only intensified this desire. Instead of selling their oil at the official posted prices, they started offering unofficial discounts and cutting the prices of their refined products. With the Suez crisis the discounts disappeared, and posted prices were increased; but as soon as the canal was re-opened this only made the disparity between the official and unofficial price levels all the greater.

The industry's one hope of avoiding serious trouble was to increase still further the rate of growth in the large markets. But at this crucial moment the United States, the most important of all, was cut off from the rest of the international industry. For several years the United States Government had been becoming increasingly restive about the growing volume of foreign oil entering

* This is the report entitled 'The International Petroleum Cartel', referred to in several earlier chapters.

the country. The defence planners argued that it would make the
United States dangerously dependent on overseas supplies in the
event of a future war, while the small producers in Texas, and
other states who had no foreign interests, claimed that they would
be pushed out of business. In 1957 the Government responded by
imposing voluntary import controls, and in 1959 President
Eisenhower made them compulsory.

This decision is one of the most important events to have
occurred in the international oil industry since the war. It meant
that all those American independents who had gone abroad to
look for new reserves to sell at home had the door to their own
market slammed in their faces. They had to look for alternative
markets overseas, and they had to do so quickly. In Venezuela
alone they had paid out over $500m. in bonuses in 1956–57, and
they had to exploit their discoveries with a minimum of delay.

The basic fact of oil-industry economics is that it costs a fortune to find and develop a field, but once that has been done, the
production costs are relatively small. The American independents
had for the most part made their basic investments before Eisenhower acted, so that a sale at almost any price was better than
none, and they naturally turned first to the big European markets.
As a result, prices in Europe dropped sharply, and in Germany, for
instance, heavy fuel oil fell from a peak of DM 142 a ton in
February 1957 to DM 60 in the second half of 1959. To add to
their problems, the major companies also had to meet growing
competition from the Soviet Union, which was trying to re-
capture its pre-war position in world markets. Russian exports
still accounted for a minute proportion of the non-Communist
world's supplies (4 per cent excluding the United States and less
than 2 per cent including the United States), but it was clear that
they could become a very disruptive influence.

The decline in the price of refined products raised political as
well as economic problems for the companies. In any ordinary
market a fall in the price of a finished product leads to a corre-
sponding reduction in the cost of the raw material. Under the
circumstances the companies should therefore have cut the price
at which their production subsidiaries sold to their distribution
organization in the consumer countries. On paper this looked a

very simple operation: all that was necessary was for British Petroleum in Germany to tell head office how much the market had fallen and for head office to tell its subsidiary in Kuwait to reduce prices accordingly. But in the oil industry life is never as simple as it looks. As a result of the tax agreements of the early 1950s, the producer governments levied their taxes on the basis of the official posted prices at which oil was exported from their countries.* They therefore refused to countenance any reduction.

Consequently, the distribution organizations in the consumer countries were obliged to pay a high price for their supplies and to resell at a low one. The companies did not mind, since their profits from production still far outweighed their losses from distribution. But the consumer governments felt the system to be unfair. It meant that the foreign-exchange cost of their oil imports was higher than the market conditions justified and that the distribution organization made heavy losses, thereby escaping taxes. Their resentment was increased by the fact that the companies frequently sold their surplus crude oil to independent distribution organizations or to countries where they did not sell refined products at a substantial unofficial discount off the posted price.

The companies sympathized with these objections, and in 1959 they tried to redress the balance by reducing their posted prices. There was, however, such a howl of protest from the producer governments that none of them was anxious to repeat the operation, despite the fact that the new level soon became hopelessly unrealistic. Then in May 1960 the Russians promised the Indian Government supplies at a price equivalent to a discount of 14 per cent off the posted price on Middle East oil. For Jersey Standard this was the last straw. On 9 August it made a further reduction with a cut of $4\frac{1}{2}$ per cent, and for a while the major companies found themselves posting several different prices for identical crude oils produced in their joint operations. This did not last long, and when the dust settled the producer governments found that the level of posted prices had declined by an average of between 2 and $5\frac{1}{2}$ per cent.

This was not a large fall, but to the governments basing their entire budgets on their revenues from oil it was an extremely

* See Chapter 14.

serious matter. Moreover, to add insult to injury the companies did not consult them beforehand. The cuts were thus a blow to their pride as well as their profits. All their latent resentments against 'foreign exploitation' came rushing to the surface in a storm of emotion, and for once they were able to sink their political differences to the point where they could act together.

On 5 September the Iraqi Government under General Abdul Kerim Kassem called a conference of oil-producing nations in Baghdad. When it opened five days later in the City Hall it was found that, despite the short notice, the delegates represented the most high-powered collection of senior officials ever to have been brought together round a single table. Perez Alfonso, once again in charge of Venezuela's oil affairs, came from Caracas, Faud Rouhani, a managing director of the National Iranian Oil Company, was there from Tehran, while the Saudi delegation was led by the formidable Sheikh Abdulla Tariki, and Kuwait's by Faisal Mezidi, one of the government directors of the Kuwait Oil Company.

As practical executives with experience of the international oil industry, these men understood why the companies had cut their prices, and some of their public statements notwithstanding, they knew that the reductions would not be rescinded. Their aim was to make unilateral action impossible in future, and to ensure that the companies always kept the producer governments' interests rather than those of anybody else in the forefront of their minds. They felt that the most effective way to do this would be to maintain a united front, so that the companies could never play one country off against another as they had done during the Abadan shut down. The result of their deliberations was the formation of the Organization of Petroleum Exporting Countries (OPEC), in effect a producers' cartel designed to hold up prices in a falling market and to harmonize the conflicting interests of its members. The wheel had turned full circle since Achnacarry, and the companies have never since tried to reduce the posted prices.

Senior oilmen often argue that their industry is quite different today from when they joined it. In a sense they are right; although the seven major companies still dominate the scene, their influence is much less than it was. To an increasing extent they have had to give way before the demands of governments, while their markets are being eroded by competition from independents and state-owned concerns. They have only managed to prosper in these unfriendly conditions because of their own astonishing efficiency, the rapid growth in the demand for oil, and the equally rapid expansion in the number of its uses.

Yet the industry's pattern today and the way it behaves are the result of its past. Much that would otherwise be inexplicable can only be understood in the light of what happened in the years between Drake's discovery in 1859 and the formation of OPEC a hundred and one years later. More than most countries the oil industry has had a continuous history, and it is only against this background that the companies' behaviour towards governments and towards each other can be understood.

PART II

How the Industry works today

CHAPTER 18

Before the revolution

'Only those who lived before the revolution know how
sweet life is.'

If the 1950s have been called the 'Golden Years of the Oil Companies', the 1960s could equally be called the Golden Years of
the Oil Consumers. As low-cost Middle East oil came to take a
more and more predominant role in oil trade outside America,
the consuming countries of the West came to enjoy a period of
unequalled expansion in supply and ever-decreasing cost. Almost
everywhere oil finally took over from coal as the fuel of industrial
expansion and almost everywhere costs of oil, in terms both of
actual prices and in comparison to other commodities, continually
fell.

Not that there were not all too obvious signs of the strains that
this expansion was putting on the structure of the world's oil
trade. The Arab–Israeli War and its aftermath in 1967 showed
that the Arab producers could, and would, put oil at the disposal
of a political cause. The dramatic outbursts of nationalism in
Dr Mossadegh's take-over of the Iranian oil industry in 1951 and
General Kassim's seizure of the non-producing concessions in
Iraq ten years later were no mere flashes in the pan. The determination to achieve national sovereignty and to exercise full control
over natural resources remained a constant theme in the debates
of the oil producing countries and the newly-formed Organization
of Petroleum Exporting Countries throughout the decade, just as
the consuming countries like Japan, Germany and Italy, without
major oil companies of their own, still looked for more national
control of their own markets and the sources of their imported
supplies.

But a radical change in the structure of the industry and the dominant position held by the international oil companies was held back by two fundamental factors dominating the industry through the decade – a continual surplus of oil and output capacity and an intensifying competition for outlets in the market-place.

Such a development was inevitable given the vast reserves of low-cost oil that had now been established in the Middle East and the framework of high posted prices based on Texas that the oil companies had attempted to maintain in the 1950s. As long as Middle East oil was still easing its way on to a world market adapted to much higher costs, the oil companies could hope to reap the full benefits of it. But as American supplies lost their pre-eminent position in world trade, as United States companies were forced to look elsewhere for a market for foreign output following United States Government restrictions on imports, and as Middle East states competed between themselves to raise production, so it became increasingly difficult for the companies to hold any agreeable price stability.

The Middle East, which had already captured the major volume of oil trade to Western Europe and Japan by the beginning of the decade, pushed forward to meet the expansion in demand and to hold its share of the market. The troubles of the 1950s over, Iranian production expanded at a rate of over 14 per cent per year from 1·2m. barrels per day in 1961 to 4·5m. barrels per day in 1971, while Saudi Arabian output expanded at a rate of about 12·5 per cent per annum from 1·4m. barrels per day to 4·5m. Abu Dhabi joined the ranks of the Middle East producers for the first time in 1962, raising its output to nearly 1m. barrels per day by the beginning of the 1970s, while Kuwait and Iraq, for both political and commercial reasons, enjoyed a rather lower rate of expansion of between 5 and 6 per cent per year, against a rise in Eastern Hemisphere demand of nearer 10 per cent.

Expansion of Middle East output alone would probably have been enough to unsettle price structures in Eastern Hemisphere oil trade, pushed as it was by governments anxious to raise their levels of oil revenues, national oil companies with crude oil of their own to sell and Japanese and European and American companies with new concessions. But the factor which more than any other accelerated developments was the growth in production from Africa, and above all, Libya. Barely on the oil map at all at the beginning of the decade, a succession of discoveries by groups such as Occidental and the Oasis consortium of Continental, Marathon, Amerada Hess and Shell shot Libya into the forefront of producers by 1970, enabling it to grow from 20,000 barrels per day in 1961 to 3·3m. barrels per day by the end of the decade – an astonishing growth rate of more than 65 per cent per annum. Nigeria, too, opened up by a partnership of Shell and BP in the late 1950s, saw a growth rate of barely less than 40 per cent per year from some 50,000 barrels per day at the beginning of the decade to more than 1·5m. barrels per day by the beginning of the next decade – most of the expansion coming after the pro-longed and bitter Civil War of the middle of the 1960s.

African crude was not necessarily cheaper than Gulf oil – indeed actual production costs averaged up to twice the 7–15 cents prevalent in the Middle East. But the oil was generally of higher quality than Gulf crudes, with a low sulphur content particularly attractive to industrialized nations anxious to reduce atmospheric pollution from fuel burning. The oil was close to the major markets of Europe at a time when the closure of the Suez Canal in 1967 made such a position all the more attractive; while, in the early years at least, the concession terms were noticeably more favourable to the oil companies than those which the companies in the Gulf had managed to obtain.

Within a space of a decade from 1960 to 1970, the share held by African oil in total West European imports rose from less than 6 per cent, much of it in the form of oil coming from Algeria to France under the France–Algerian connection, to more than 40 per cent – a rise involving a fortyfold volume increase from around 200,000 barrels per day to nearly 5m. barrels per day. The delivered cost of Libyan oil especially almost consistently undercut the cost of oils from the Gulf, largely because of its freight advantages. And this in turn helped to bring down the average price of landed Gulf crudes.

But it was less the effect of North African oils on crude oil prices – a development which was partially restrained by OPEC's effective action in enforcing a stabilization of posted prices in Africa as well as the Gulf – than their impact on the market itself which was most strongly felt in the oil industry. Libya in particular for the first time gave a number of independent companies an extensive source of oil of their own, either to sell themselves in the market through their own refineries and marketing systems, as Occidental and Continental Oil developed, or to sell to others, as Amerada Hess, Marathon and other companies did.

The fact of availability of crude oil in turn encouraged companies in Europe to enter into particular aspects of the oil trade such as refining or petrol marketing, where they could take advantage of surplus oil at the margin and the particular advantages offered by high growth markets such as urban petrol sales. Without the overheads that the majors had to carry in order to maintain a continuous flow of oil through all phases of the business as well as a widespread international network of sales of all products, the independents were able to step in to 'cream off' the most lucrative parts of that trade, often on the basis of cheap oil picked up from the surplus markets.

The response of the major oil companies was to try to make maximum use of their size and integration to lower costs. As the decade wore on, more and more investment was poured into the creation of larger refining units, bigger tankers, larger terminals, combined pipeline distribution systems and high-volume retail outlets in order to lower unit costs and take advantage of growth. Whereas in 1960, two-thirds of the world's tanker fleet was made up of ships between 2,000 and 30,000 dwt and none was larger than 90,000 tons, by the end of the decade less than 20 per cent were in the 2,000–30,000 dwt class and nearly a third were 100,000 dwt and over – a move that was helped by the need to keep Gulf freight costs low in view of the competition from African oils and the eventual closure of the Suez Canal. The same could be said of refining, where the capacity of the basic distillation units went up from one or two million tons a year to new units as large as 9m. tons at the end of the 1960s; and of distribution and marketing facilities, where unit sizes of storage, road tankers and other facilities all showed the same development towards economies of scale.

But this trend in turn only added further fuel to the fires of competition. The greater the pressure towards increasing scale of operations the greater the temptation of the majors themselves to look for a higher market share to justify the costs. The financial attractions of large-scale units really become apparent when the 'load factor' exceeds 80 per cent or more and the natural tendency therefore is to treat the margin above this as the real profit earner and to seek outlets for it virtually regardless of cost.

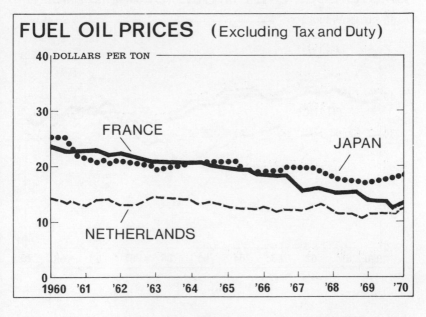

FIG. 5

The effect on the market of this pursuit of marginal sales was dramatic and increasingly self-destructive. In a few countries like France, where government direction of the market deliberately constrained competition, and in some other areas like the United Kingdom and Japan, where the majors held their predominance in the market until fairly late on in the decade, the oil companies managed to retain some semblance of price stability. But in Africa and many parts of the developing world, their position was continually eroded by an increasing trend towards nationalization of marketing facilities and by the pressure from governments on companies to build large refineries. Meanwhile, in the great mass of the European markets, particularly Germany and Benelux, prices throughout the market became more and more influenced by those in the marginal open markets of Rotterdam and Italy. At the upper end, petrol became the scene of

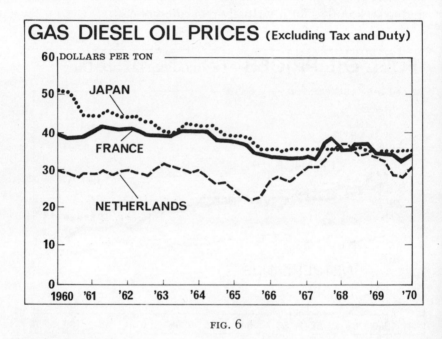

FIG. 6

fierce competition and price-cutting led by independent operators concentrating their attention on limited geographical representation in the fast-growing urban centres. At the other end, fuel oil became increasingly a 'distress product', the subject of heavy competition between the majors and more and more influenced by the yardstick of prices set in Rotterdam.

During the 1960s, bulk prices of most oil products in north-west Europe fell sharply early on, rose again during the 1967–68 crisis of the Arab–Israeli war before collapsing again in 1969, and the unit proceeds from the market of most of the major international majors followed roughly the same pattern. To a surprising degree, indeed, oil company profits became dependent on freight rate movements – the one variable cost in their integrated chain.

When freight rates were low – as they were for most of the decade – the cost of marginal crude itself tended to be low and the oil companies, with their mixture of owned tankers and tankers on long- and short-term charter, tended to suffer. When freight rates soared, as they did during the 1967–68 crisis, then the oil companies with the same 'cover' on tankers tended to benefit, while buyers of oil on the open or 'spot' markets were forced to pay high prices for the marginal oil. For many analysts among stock brokers, the question of predicting oil company returns became a matter of judging freight rates and determining how well off any particular company was as far as its need to go into the market for short-term charters was concerned.

This pattern of increasing scale of operations coupled with rising price competition was not unique to the oil industry. On the contrary, it marked the development of a whole range of industries during the 1960s from retailing to aviation and chemicals. But what made the oil industry unique was the continuing dominance of the major oil companies over the source of the raw material and the determination of the producing governments to ensure that competition did not erode their own revenues which were based on a percentage of an official, or posted, price of oil. Crude oil prices did decline through the adoption of rising rebates on the official prices in arms-length deals between companies. But the real erosion took place at the market end, where competition was fiercest and the cost of marginal oil supplies had their greatest impact.

For the consumer, this was pleasing enough. He got all the oil he wanted at what were, by the standards of any other commodity and certainly by the standards of any other fuel, bargain prices. Some commentators, such as Professor A. Adelman in the United States and Dr Paul Frankel in the United Kingdom, argued – with considerable force – that the benefits to the consumer were still not all they should be.

The system of host government taxes based on artificially high posted prices agreed between the producers and the oil companies, it was pointed out, maintained a price floor of tax-paid cost on oil well above its true value calculated by any normal standard of supply and demand. Remove that floor and the oil companies which acquiesced in it and prices would fall to a natural level of the cost of bringing in additional or alternative supplies.

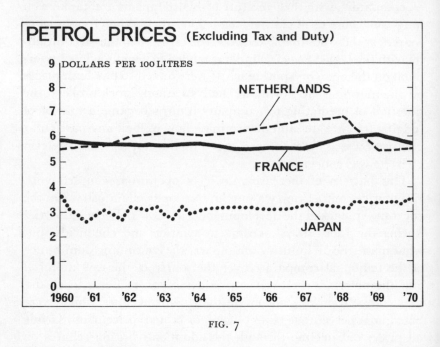

FIG. 7

Most oil consuming governments, however, did not see the situation in this light. Oil was relatively cheap and getting cheaper. In some sectors, such as transport, there was no substitute for it and almost every consuming nation, far from seeking to lower the price, actually raised it by putting on progressively more and more excise tax, largely for reasons of easy revenue-raising. On petrol alone during the 1960s, as much as 75 per cent of the final price of petrol was made up in consumer taxes, which varied from around 32 per cent in the United States, around 58 per cent in Japan and between 58 and 76 per cent in Europe. Consumer government taxes on gas oil, though less in cash figures, accounted for as high a proportion of the selling price while even fuel oil was taxed at a rate of between 15 per cent and 33 per cent of the final selling price in Europe, although not in Japan and the United States.

If consumer governments were worried about developments in the oil trade it was more on the basis that oil was threatening jobs in the indigenous coal industry (as in the United Kingdom) or that the grip of the foreign-owned international oil companies presented potential dangers of security of supply (as officials in France and Italy argued). And neither of these lines of argument promoted the thought of cheaper prices – in fact the opposite.

For the oil consumer to have broken through on the posted price question would have required a political cohesion amongst consuming governments coupled with a readiness to confront the producers that was quite obviously lacking at the time. Equally, for the consuming governments to have attempted to by-pass the major oil companies and go for direct relations with the producers, as ENI of Italy urged, would have needed a general dissatisfaction with the way things were going and a conviction that direct deals did in fact provide greater security of supply – feelings that were far from apparent amongst the majority of consumers at the time.

But, for the same reasons that some consumers urged the withdrawal of the major oil companies from their producing role in order to achieve lower prices, the producers themselves found it easier to co-exist with them. As long as there was a surplus in the market and competition for outlets for crude oil, the major oil companies provided a stable means of ensuring those outlets – an institution sufficiently dependent on Middle East supplies to be bullied on price and tax questions and a necessary source of the technical and management expertise to handle production and distribution of the crude oil.

In the creation of the Organization of Petroleum Exporting Countries* at the beginning of the decade, the producers had at least presented a sufficiently strong co-operative front to prevent a fall in posted prices in line with the market price. And, to a certain extent, it can be argued that the very fact of producer unity made it not only more difficult for the companies to refuse their demands but actually made it in the companies' interests

* The founding members of OPEC, Iran, Iraq, Saudi Arabia, Kuwait and Venezuela, were joined by Qatar in 1961; Indonesia and Libya in 1962; Abu Dhabi in 1967; Algeria in 1969; Nigeria in 1971 and Ecuador in 1973.

to accept them. The great danger from the companies' point of view was the prospect of individual action in which a company might find itself saddled with higher costs than its competitors. So long as the terms were applied evenly throughout the producing world, then the competitive problems were less and might actually be in favour of the major oil companies if it helped to reduce some of the competitive advantages previously held by the independent producers in Libya where the original concession terms were much lighter on the companies than those prevailing in the Gulf.

Yet to imply collusion between the majors and the producers would be to seriously underestimate both the difficulties faced by OPEC in its struggle to gain better terms as well as the resistance put up by the companies to its demands. From the producers' point of view, the posted-price system was the basis of the oil revenues on which they depended for economic development. Understandably, they felt that the exploitation of their main natural asset should be conducted in a way that would serve their own internal needs. For the oil companies, on the other hand, the posted-price system could all too easily become a method under which the oil producers' revenue per barrel remained constant while their own share in the profits, because actual market prices were falling well below posted levels, declined.

The difficulties became apparent from the start. The producers, having formed an administrative structure for co-ordinating their interests in OPEC, then moved to tackle the most immediate problems. In 1962 they sought the triple task of raising revenues by eliminating the market allowances traditionally given to companies as a percentage of prices, of tackling the important issue of 'expensing royalties' and finally of pursuing their major ambition of restoring posted prices to the levels before the cuts introduced in 1960.

The market allowance issue was fairly speedily settled in negotiations between the Saudi Arabian government and the Aramco consortium the same year. But the issue of expensing of royalties was a different matter. Traditionally the oil companies had treated royalties as a sort of partial payment of revenue dues, deducting them in full from the 50 per cent tax assessment. The producers, however, argued – with considerable justice – that royalties should be treated only as an expense before the tax calculation. The potential difference to government 'take' per barrel was about 11 cents, or around 15 per cent, and negotiations dragged on for three years before agreement was reached.

The oil companies refused to accept OPEC's right of collective negotiation on behalf of the producers as a whole and would only discuss the issue on an individual country-operating company basis, insisting on a discount off posted prices as the *quid pro quo* for accepting royalty expensing. The producer countries themselves became divided over the issue of just how far to compromise on their agreed demands. When a broad settlement was finally reached at the end of 1964, it was far from being an outright victory for OPEC. The oil companies were able to obtain a discount off posted prices, to be progressively reduced over its three years of operation. But, in shades of things to come, Iraq refused totally to accept the compromise. The Kuwaiti government found itself in unexpected difficulties in gaining National Assembly approval to the deal and it took a dubious election of a new and more pliant Assembly before the agreement was ratified nearly three years later. Libya meanwhile took the opportunity of wrapping up the royalty expensing issue with other tax questions and it was not until after a bitter fight with the threat of expropriation and export controls that agreement with all the operating companies was finally achieved in 1966.

The basic problem for the producers was that, while OPEC had become a useful means of establishing certain minimum bargaining points with the oil companies, it could only become a really effective organization to control prices if its member states would agree not to compete amongst themselves for production increases – the classic monopoly stance. That it was not able to do this became apparent as soon as OPEC tackled the problem of trying to restore posted prices to their former levels before the 1960 cuts.

It was for this purpose that OPEC had been created and from very early on there were suggestions that it should set production rates for all its members in order to remove the surplus of oil in the market. The strongest proponent for this scheme was Venezuela, whose position in world oil trade had been most undermined by the tremendous expansion of Middle East output. But it was an argument hardly likely to appeal to the newcomers to production like Libya and Abu Dhabi, who were just starting off on the road to oil wealth, any more than it was to countries like Iran and Saudi Arabia, who saw in rising production the clearest route to ever-increasing revenues.

Having resolved to tackle the issue in the first OPEC conferences in 1960, it was allowed to fall into the background until, under the constant urging of Venezuela, it was revived in 1965–66. The Organization's new economic commission was set the task of putting up a rational programme of production expansion over the coming year with each country being set a 'quota'. Although grudgingly accepted by the member states on an experimental basis, the project never really got off the ground. Countries quickly took the quotas not as a ceiling of output but as a minimum output rate to be sought, complaining bitterly when it was not achieved and equally only too happy when they found that they had met or exceeded their 'allowables'. Far from achieving its aims, the issue only raised greater dissent between OPEC's members and was allowed quietly to fade away.

Without the achievement of a really effective cartel in OPEC, much of the initiative was left to individual producers. The later 1960s became a period of almost constant negotiations between operating companies and their host governments over production rates, exploration activity, the release of non-producing acreage and tax questions. Iran, in particular, became increasingly ambitious to raise her revenue through higher output to increase her rate of internal economic development – a direct relationship between oil production and national need which the international oil companies, seeking maximum flexibility to suit marketing conditions, were reluctant to concede. And as Iran pushed the operating consortium there into a series of production expansion agreements, culminating in the 1969 agreement to provide the revenue it required through higher output and loans, so other countries like Saudi Arabia and Kuwait pushed the oil groups in their lands to gain similar treatment.

When it came to the crunch, however, as in Iraq's prolonged disputes with the Iraq Petroleum Company, the bitter dispute between Syria and IPC over pipeline dues in 1966 and the Arab–Israeli War of 1967, the unity in the producing world was not sufficient to enable the more extreme calls for nationalization or the extensive use of oil as a political weapon to gain a real hearing.

Development in Iraq, with the failure of the government to ratify an agreement settling outstanding differences in 1965, tended to stagnate as the two sides accepted a virtual stalemate position. Libya meanwhile managed to clear up a number of anomalies in its tax and concession structure but failed to really break out of the legal restrictions which bound it to the original concession clauses on repatriation of profits, 'most favoured company' treatment and arbitration procedures. Saudi Arabia enjoyed steadier growth following the deposition of Ibn Saud and the resignation of Abdullah Tariki from the post of Oil Minister. Other countries in the Gulf were more content to share in the achievements of others while Kuwait, its pre-eminence in the Middle East output stakes gradually eroded over the decade, was perhaps too nervous of its position of wealth amidst stronger neighbours and too uncertain in its own internal political scene to push issues to their limit.

By the end of the 1960s, therefore, many of the forces which had appeared a decade previously were indeed taking shape. The producing countries had formulated their ambitions to control production in the interests of their own political and economic needs. In OPEC they had at least fought a successful defensive action against the depression of posted prices and had actually increased their average revenue per barrel through the reduction of marketing allowances and the expensing of royalties. In individual negotiations they had managed to extract compromises on a number of issues of taxation, return of concession acreage and output rates. But they had yet to achieve the full control of the exploitation of natural resources that the more radical spokesmen had wanted any more than the consumers themselves had managed to take over control at their own end.

The oil companies, meanwhile, had found themselves caught between the pressure of market competition at one end and the ambitions of the producers at the other. To a surprising degree the major oil companies still retained a predominant position in oil production and a majority share of refining and transporting as well, but their return on net assets fell from an average of over 15 per cent in the 1950s to around 11–12 per cent in the 1960s and an even greater decline on their returns in the market.

Although to some extent buttressed by the continuing profitability of American operations, where the return on net worth amongst the major oil companies actually increased from over 11 to more than 12 per cent in the middle part of the decade, average returns in the Eastern Hemisphere declined from over 13 per cent to barely more than 11 per cent at the end of the decade, while earnings per barrel were reduced from nearly 55 cents to around 33–35 cents by 1970. Even assuming that these figures exaggerated the real suffering, the fact was that the once most glamorous sector of western industry had fallen back by the end of the 1960s to a level of profitability which was little better, if not actually worse, than that enjoyed by manufacturing industry as a whole.

Looking to the 1970s, there seemed little doubt that oil's phenomenal growth in the world's energy pattern would continue unabated. The central questions were whether the oil industry could sustain the level of investment needed for that growth and where the balance of power between the producers and the oil companies would finally settle itself. Worries there were in plenty, but for most it was likely to be a question of more of the same thing than anything dramatically new.

CHAPTER 19

The revolution

'In looking back on the Tehran and Tripoli negotiations of 1971 with the benefit of hindsight, I would characterize them as having been as successful as the circumstances at that time allowed.'

George Piercy of Exxon

'Our policy and negotiations since 1971 have been an unmitigated disaster.'

Henry Schuler of Hunt International Petroleum*

The fundamental confidence in demand growth and supply stability engendered in the world's oil trade during the 1960s was rudely shattered at the very opening of the next decade. An oil industry increasingly geared to conditions and prices at the margin was suddenly upset by a rapid reversal in marginal conditions. Higher demand than expected in 1970; a shortage of tanker tonnage; a prolonged closure of the Tapline pipe carrying Saudi Arabian oil to the Mediterranean and a growing tightness in European refinery capacity all combined unexpectedly to throw the world's oil trade out of gear and wipe away the marginal surpluses of products which had dominated the previous decade.

An industry which had prided itself on its flexibility and its ability to overcome the supply crises of successive closures of the Suez Canal was now revealed to have been left with very little real flexibility as demand pushed hard against supply. The consuming areas of the West, which had based their energy policies on continued availability of cheap oil over the future, now found

* Statements made before the Senate Foreign Relations Subcommittee on multinational companies, 1974.

themselves uncomfortably unable to do without even relatively small amounts of that oil. The producing countries, who had spent a generation attempting to sustain a reasonable amount of revenue against adverse market conditions, now found themselves with the power not only to resist market trends but to control them.

Ironically enough it was Libya, the country which had been most responsible for the surplus conditions in the market during the 1960s, which was now to lead the battle in changing those conditions and displaying the potential power of the producers under the new conditions. The development of oil in Libya was perhaps never the most glorious chapter in the history of the oil industry. Against a background of a weak and inexperienced government under the rule of King Idris and a world picture in which the low cost of Middle Eastern oil predominated, the oil companies had been able to extract highly favourable conditions from Libya during the late 1950s and early 1960s. Tax dues and royalties were based not on straight posted price but on posted prices less 'marketing allowances' – an allowance which some of the operating companies were not slow to take advantage of. Concession terms were protected by 'most favoured company' clauses ensuring that all concessions gained the benefit of the most favourable terms going; by clauses ensuring that the government could not introduce laws affecting concession terms without the consent of the operators and by clauses enabling the companies to repatriate funds freely.

As Libya became a major oil exporter in its own right, and as it joined forces with the other leading producers in the Organization of Petroleum Exporting Countries, dissatisfaction with these terms on the part of the government, now aided by a number of foreign advisers, became more and more apparent. The royalty expensing issue provided the opportunity, and in 1965 the government passed a new law attempting to establish tax terms on the same basis of posted prices as already prevailed in the Middle East. The majors, who had negotiated with OPEC in the Gulf and had no desire to see the independent companies in Libya get away with better terms and thus cheaper oil, complied. The independent companies, which accounted for more than

9. (*a*) Mossadegh, the revolutionary who failed

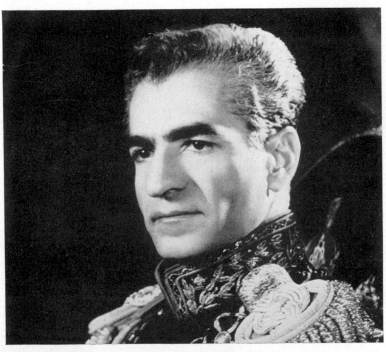

9. (*b*) The Shah, the gradualist who succeeded

10. (a) Disposing of natural gas: flaring it off in the Libyan desert

10. (b) Disposing of natural gas: edible protein in a British laboratory

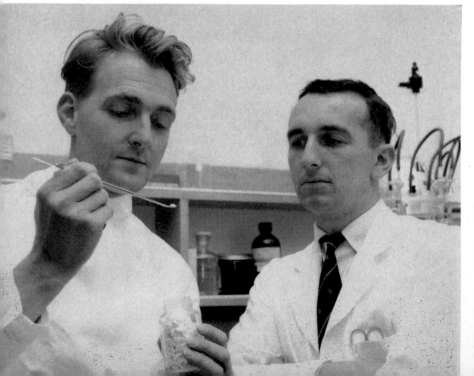

half total Libyan output and who feared that new terms would enable the majors to undercut them with cheaper-cost Gulf oil, refused and it was only after a bitter struggle, fought partly in the world's press, and the final threat that the Government might take action in prohibiting exports or actually nationalizing assets, that all the companies fell in line in 1966.

The new law, coupled with moves to tighten up government supervision of company accounting and to accelerate the payment of tax dues, enabled Libya to achieve one of the highest revenues per barrel anywhere in the Middle East. Within a few years, however, the closure of the Suez Canal in 1967 had again upset conditions. Libya, like Algeria beside it, gained increased advantages from its proximity to the market and hence lower shipping costs to the companies. Its oil, low in sulphur and with a higher natural yield of petrol and heating oils in the refining process, was also becoming increasingly valuable in a market where petrol consumption was rising rapidly and where consuming governments were showing more and more consciousness about pollution problems. By early 1969, both Libya and Algeria were pressing for higher prices to reflect this greater market value and were accompanying these demands with claims for back dues on the same score.

Before talks between the Libyan government and the companies could start, however, the régime of King Idris was suddenly and unexpectedly overthrown by a group of young officers led by Colonel Ma'ammar al-Qadhafi. Committed to the ideals of socialist revolution in the image of Nasser and Islamic fundamentalism, the new régime was bound to see in oil a major battleground against Western, United States-dominated neo-colonialism. They were in close contact with Algeria, which was all too eager to use Libya as an additional arm in its struggle to revise the terms of the 1965 Franco–Algerian agreement.

Where the previous régime, and particularly the King, tended to step back from final confrontation with the oil companies, the new régime had few qualms on this score, and its taste for battle with the West was only increased by the ease with which it was able to get the United States and Britain to evacuate their military bases in the country within a few months.

Even so, the oil issue did not immediately break out into open conflict between the companies and the new government. The Revolutionary Command Council re-affirmed the country's intention of pursuing higher prices and, in opening negotiations on the subject at the beginning of the following year, made it clear that nationalization was not on its mind, that it regarded, in public at least, good relations with the oil companies as essential but that it was determined to ensure that the Libyan people gained their 'fair demands and legitimate rights'. The obvious worry of the oil companies was that any concessions made to Libyan demands might upset their relations with producers elsewhere and undermine the general price structure prevailing through the Middle East. But the new oil minister, Izz al-Din Mabruk, assured them that it was not Libya's intention to destroy this general framework.

Negotiations thus dragged on through the first half of 1970 with surprisingly little public barracking. Where the previous Libyan régime had suggested that it was looking for a rise of around 10 cents on the $2.20 posted price for its oil, the new régime was less specific but clearly more ambitious. In reply, the oil companies – in this case mainly Esso and Occidental – offered around 6 cents, which was promptly rejected. But, although the talks were far from completely fruitful, the feeling even as late as April and early May was that a compromise could and would be reached, possibly in the region of 12 cents per barrel.

It was not to be. By summer, Libya's determination to gain really significant price changes for its crude had been greatly strengthened by an agreement to co-ordinate its efforts in this direction with Algeria, which was still pressing for much higher posted prices from France, and with Iraq, whose disputes with the Iraq Petroleum Company still formed a running sore in oil industry relations there.

At the same time market conditions were swinging more and more in favour of the short-haul oil suppliers to Europe. The years of marginal pricing and poor returns had had their effect on investment. Tanker tonnage was becoming in short supply with the continued closure of the Suez Canal. The general resurgence of economic growth in virtually every part of the industrialized world at the same time raised demand not only for oil shipments but for dry cargo trade as well. And this lack of surplus tanker capacity was only made tighter by the sudden loss of several of the new class of 'mammoth tankers' in mysterious explosions early in the year and by the rapid rise in demand for heavier oils to provide fuel oil to European industry and power stations, which in turn resulted in a stronger pull on long-haul supplies of heavier crude oils in the Gulf. Refinery capacity in Europe began to look as if it might be strained for the first time in more than a decade and, in the most dramatic blow of all, the Tapline pipe taking nearly half a million barrels per day of Saudi Arabian production to the Mediterranean port of Sidon in Lebanon was damaged by a bulldozer in May and the Syrian authorities refused to allow any repairs to be made. It was not repaired until the beginning of the following year.

The implications of these developments on their own cause was not lost upon the Libyans and Algerians, who now began to raise their expectations and their demands. And then, in May, the Libyan authorities began to implement the most effective squeeze of all on the oil industry – progressive cut-backs to company oil production on the grounds of conservation. Whether they did this from the beginning in full consciousness of its effect on the price talks or whether it was the product of genuine concern about preserving a reasonable rate of reserve depletion remains unclear. Certainly the conservation issue and the accusations that some of the companies were depleting fields too rapidly had been a consistent subject of Libyan concern ever since the mid-1960s and had become a central tenet in the oil policy of the new régime from the moment it gained power.

But, either way, the Libyan government soon understood and pursued conservation cut-backs as a devastating weapon in the price negotiations. Starting from a relatively small reduction in Occidental's output in May, the government successively imposed cuts of much greater significance on almost every major producing group until total Libyan output was down by over 1m. barrels per day on previous rates. On top of this, the Libyans also began to change negotiating tactics and concentrate their attentions not on Esso as before but on Occidental and the independent operators in Libya. Where, in previous disputes between the companies and the Libyan government, it had been the majors which had proved the more tractable and the independents which had proved the tougher in negotiations, now it was the other way round.

The majors, while still being extremely vulnerable to any loss of short-haul crude oil at a time of such tightness in supply, were clearly worried about the effects of any deal made in Libya on their much more extensive holdings in the Gulf area. The independents, like Occidental and Continental Oil, on the other hand, were virtually entirely dependent on Libyan oil for their Eastern Hemisphere supplies. There was little love lost between them and the majors, and few, if any, had holdings outside Libya to worry about.

The combined effect of progressive cut-backs to production and the concentration of pressure on the independents was finally overwhelming. In early September, Occidental, whose production had been cut from over 800,000 barrels per day to barely more than 400,000 barrels per day, signed an agreement giving Libya an immediate rise of 30 cents per barrel on its posted price rising to 40 cents in annual increments of 2 cents over the following five years coupled with an acceptance of a 5 per cent rise in tax rates in back payment of what the Libyans claimed should have been higher postings since 1965.

Once Occidental signed, the other independents quickly followed. But Shell, a member of the Oasis consortium with three United States independents, stood out on the principle that such a retrospective payment would 'sell the pass' to any sort of claim by any producing government and was totally unjustified in terms of the changed market conditions between 1970 and previous years. In this it was informally supported by BP and, to some extent, by Esso.

Whatever the logic of its argument, Shell's stand proved ineffective in the event. The majors were too weak and too divided to undertake a concerted action over Libyan prices. Most of them were desperately worried about their own financial results during the year and fearful of the effect that continued loss of Libyan production might have. The consuming countries were unwilling to back them in a struggle that might endanger supplies and Libya itself was openly threatening nationalization if it did not get its way. By the end of September, Texaco had quietly settled. The other majors unilaterally raised their posted prices to preserve the appearance of still being masters of prices but eventually give in to the retrospective tax terms as well. In December, Shell, the last holdout, finally signed on the dotted line, agreeing terms that were identical to those which Occidental had accepted.

It was a famous victory, indeed. While Texaco, in breaking ranks with the other majors, had argued that a settlement in Libya need not upset agreements in the Gulf, the event was far too important in its implications to be localized in this way. Libya had shown that the oil companies and the consumers needed the producer far more than the other way round particularly for a country like Libya whose revenues were far greater than her needs. It had achieved not just a new agreement with the oil companies but a complete rout of them. It had broken the formal 50/50 profit-sharing arrangement that had been the basis of oil relations for the previous twenty years. It had established the producer countries' power, if not right, to raise posted prices for the first time and it had forced the companies to accept the unacceptable principle of retro-activity on prices. Most important of all, it had displayed the extent of power which the producers now held in a world inescapably dependent on their oil where surpluses were no longer a *de facto* part of life.

The majors, only too aware of the problems that this was likely to bring to their relations in the Gulf, moved quickly to offer the Middle East producers new price and tax terms to hold the line. A settlement was reached with Iraq, Saudi Arabia and Nigeria giving them similar terms to Libya's on its short-haul crude oil. Iran in November accepted a 5 per cent increase in its tax rate and an offer of a further 9 cents per barrel on the heavier crudes which were now in high demand because of the demand for industrial fuel oils in Europe and Japan, and then the same offer was made to all the other Gulf producers.

Whether the oil companies really thought that this would stabilize the situation is doubtful. Panic was beginning to set in at the market. Freight rates were rocketing to record levels.* There were widespread fears of actual shortages occurring if cold weather raised demand much further and the producers had only to read the newspapers to see that conditions had fundamentally changed in their favour and that they still had a long way to go before reaching the price ceiling for oil.

Meeting in Caracas, Venezuela, in mid-December, OPEC quickly sensed that here at last was the moment to press their long-felt ambitions after the frustrations of the 1960s. Deciding on a unified strategy on prices, in which talks would be conducted on a regional basis of the Gulf countries, the Mediterranean exporters and Venezuela–Indonesia, the Organization called for higher prices, the establishment of a minimum income tax rate of 55 per cent and elimination of previous allowances against tax for gravity and marketing expenses, and established a tight schedule for negotiation with the threat of unilateral legislation of their demands if they were not met.

This massing of producer forces was daunting enough for the oil companies and the consumers, threatening as it did a massive jump in oil prices for the first time almost since the Second World War. What was even more disturbing was the possibility that it could lead to a rash of demands from producers in various parts of the world which promote a never-ending spiral of leap-frogging demands around the world.

The deepest fears seemed only to be confirmed by the weeks that followed the Caracas meeting. Venezuela at the turn of the year passed laws raising the tax rate from 52 to 60 per cent and giving the government the right to raise unilaterally its tax-reference prices. The Franco–Algerian talks bogged down in Algerian demands for control and massive price increases. Worst of all, Libya suddenly returned to the fray in early January with a

* Some idea of the extent of the panic, and the power that the short-haul producers like Libya held in the market at the time, can be gained from the fact that at the height of the crisis in January 1971 freight rates for a Persian Gulf–North West Europe tanker trip reached Worldscale 270, the equivalent of $24.75 per ton, compared with an average of around $2.50 per ton through the 1960s. To replace one tanker of Libyan oil meant the chartering of nearly three tankers of Gulf oil because of the greater distances involved.

statement that its previous agreement had been solely in the form of a settlement of outstanding views, that it had not gained the benefit of the most recent changes in market conditions and that therefore it was presenting 'non-negotiable' demands for a further 5 per cent rise in tax rates (on the ground that the previous increase had been in lieu of retrospective payments) and for a further sharp rise in prices to reflect the higher freight rates prevailing in the market.

For almost the first time the consuming nations as a whole were brought directly into the price-bargaining process. Representatives of the United States' State Department, France, Britain and the Netherlands, as the countries whose oil companies were most directly involved in the discussions, met in Washington through the first weeks of January, consulting with the other producing areas such as Japan as well. This was followed up by an intense diplomatic effort through ambassadors in the various producing countries and a visit to the Gulf of a senior official of the United States' State Department, John Irwin, to get the producing countries to withdraw their threats to cut off oil supplies and agree to stop leap-frogging.

At the same time the oil companies, spurred into urgent action more than anything by the size of the Libyan demands and pressed hard in the Gulf by the series of forceful statements made by the Shah of Iran and other leaders in response to the dilatory start to Gulf talks on 12 January, met hurriedly in New York to decide what they could do. The independent companies, for so long the rivals of the majors, were now intensely concerned about the threat to their position in Libya, and anxious for some kind of support from the majors should the worst come to the worst. The majors were equally anxious that the independent companies, who had already been called in to start talks in Libya, should not sell the pass once more and set off a new round of leap-frogging demands between the short-haul and long-haul producers.

With United States' Justice Department permission and with the general backing of the consuming governments, the companies decided to join their forces together, to guarantee to support each other's position and to seek to negotiate as a co-ordinated group. The result was the famous letter of 16 January, delivered to OPEC and signed by thirteen oil companies including all the majors, CFP of France, and almost all the United States independents operating in Libya.* The letter offered the producers a new five-year price plan, including an immediate increase in prices and annual adjustments against inflation; the principle of a freight premium for short-haul producers and the recognition of OPEC's right to negotiate on behalf of all producers in return for guarantees specifically aimed against Libyan demands that there would be no further increases in tax rate, no separate negotiations with individual producers and, most important of all, that any settlement should be on the basis that it be reached 'simultaneously with all producing governments concerned'.

Thus was set the stage for what, in formal terms at least, was to be a battle of the giants. On the one side were the massed ranks of the producing countries claiming – not without a good deal of justice on their side – that oil had been underpriced in the market compared to its true value as set against the cost of other fuels, that market forces and the power of the oil companies had prevented the producers gaining the full share in the exploitation of its resources and that prices in the market should reflect more of the producing countries' own needs for revenue. On the other side were the combined phalanxes of the international oil companies, partially backed by the consuming governments arguing that oil prices should reflect market conditions of supply and demand, that a dramatic price increase could do much harm to the economies not only of the West but also of the developing world and that oil was far too important a commodity in world trade to become a pawn in the competing ambitions of the producing world. Behind this conflict was the fundamental question which had been at the bottom of producer–oil company relations for the past generation – the question of whether oil should be developed as any other commodity in line with consumer needs, cost of production and availability of supply or whether it should

---

* The original thirteen were eventually joined by eleven others, including Hispanoil of Spain, Gelsenberg of Germany and Japanese interests. The two European state oil companies, Elf–ERAP of France and ENI of Italy, held back, however, on the ground that their interests were 'different' from those of the majors.

be developed in accordance with the needs and ambitions of the individual producing states acting in combine to enforce their will.

In reality the particular negotiations of 1971 proved considerably less high-minded than this. Both the consumers and the oil companies were less concerned about a climactic showdown on the price question than the urgent need to achieve some stability of pricing and supply within which future growth could take place. Price was still not a large enough factor in the costs of the industrialized countries to promote a confrontation on the question while the oil companies, so long as they could pass on any increased prices to the consumer, lacked the final self-interest needed to have taken this issue to the point of self-destruction.

The producing countries, on the other side, had neither the cohesion nor probably the will effectively to use a cartel to drive prices to their absolute maximum. There were clear differences between member states, particularly when it came to political matters or questions of changes in the structure of the industry. There were even more striking divisions between the short-haul producers and the Gulf oil states. The primary aim of the Gulf states was to achieve a substantial rise in prices, and the oil industry's letter of 16 January was thus greeted with a cautious welcome in the Middle East as containing at least a recognition of their case for higher prices. In the Mediterranean, on the other hand, Algeria and Libya had much more ambitious plans for a victory over the oil companies and thus bitterly attacked the oil industry initiative as an attempt to dictate the ground-rules of discussion and divide the producers.

Predictably, the oil industry's initial attempt to get negotiations started on a global rather than a regional basis failed from the outset. Despite some intense political activity on the part of the United States' State Department and the efforts of the companies themselves, the Gulf states refused outright to accept any connection between negotiations with their representatives and the negotiations in the Mediterranean, nor would they agree to regard themselves as responsible or tied to agreements that might be reached outside the Gulf. The London Policy Group, set up along with a similar body in New York to co-ordinate the companies' stance in the negotiations, within a few days split into two groups, one for the Mediterranean and one for the Gulf, and two negotiating teams. A Gulf team headed by Lord Strathalmond of BP and a Mediterranean team headed by George Piercy of Esso, were briefed with an offer to the producers.

Still seeking to maintain an appearance of parity between the two sets of talks, the company offer was couched in the same terms, with the same provisos about no further leap-frogging and with financial terms that were strictly related to each other. While Lord Strathalmond's offer was greeted as a reasonable start to Gulf negotiations, however, George Piercy's visit to Libya proved little short of farcical. The Libyan government absolutely refused to recognize the right of the negotiating team to talk on behalf of the industry as a whole, insisting that discussions could only be on an individual company-to-government basis, and the oil company delegation was finally forced to resort virtually to slipping its offer underneath the Oil Ministry door in Tripoli where it was steadfastly ignored. However much the oil companies attempted to preserve the fiction of parallel and inter-related talks, the fact was that the two sets of negotiations had become separate issues in which the Gulf would set the pace on their own behalf and the short-haul producers would wait and see what they could do as a follow-on.

The Gulf talks, although accompanied by a more gentlemanly atmosphere, were far from easy. The negotiations began on 19 January under a strict deadline set by OPEC of 3 February and, from the start, the gap between the two sides was wide. The producers demanded an immediate rise of 49 cents per barrel, going up to an 87-cent rise by 1975 and involving a total increase in producer country revenues over the period of some $12,000m., or well over 50 per cent. The companies' offer was for a far more modest rise of an immediate 15 cents in the posted price rising by degrees to 22 cents by 1975 – a total increase of just over $3,000m. over the five years. Although this was later amended to an immediate rise of 27 cents per barrel rising to 43 cents by 1975, it was still only what the Gulf states were calling for.

Added to this, there was the continuing importance of the guarantees against leap-frogging between the Gulf and the

Mediterranean – a point of vital importance to the oil companies but one on which the Gulf states refused to commit themselves. It was in an atmosphere of very considerable tension therefore that talks broke down altogether on 2 February and a full OPEC meeting was called to decide what was to be done.

Despite the drama of this occasion, when the Gulf members of OPEC threatened to legislate their demands if their minimum price requests were not met by 15 February, and when the Shah of Iran threw in his full prestige behind the talks, it was more of a final display of producer government strength before the inevitable victory than a real plunge into a confrontation, which neither the producers, consumers nor oil industry wanted. On the price question, there was little real prospect of either the companies or the consumers standing up to the united forces of the producers, particularly when it was accompanied with threats that they would simply introduce new price structures unilaterally. But there were already signs of a compromise emerging and when the producers spoke of 'minimum demands' it was for a sum of nearer 35 cents per barrel than 50 cents. At the same time, too, it was becoming clear that the producers, while unwilling to try to influence the Mediterranean talks, were not willing either to support the demands of the short-haul producers. The OPEC meeting saw some bitter comments being passed between Libya and Algeria on the one side and Iran and Saudi Arabia on the other and, although the final resolutions did express broad support for the short-haul cause, they stopped far short of an absolute commitment by the Gulf states to support their colleagues in the Mediterranean whatever their demands.

A private meeting between Lord Strathalmond and Dr Amouzegar took place in Paris. Negotiations restarted in Tehran on 12 February and by 14 February there was agreement, giving the Gulf states an immediate rise of 35 cents per barrel, re-adjustments on gravity differentials and annual increases of 5 cents per barrel plus 2·5 per cent of posted prices over a five-year period. In return, the companies gained firm guarantees against leap-frogging in the Gulf, the promise of stability for the next five years and an assurance that the Gulf producers would not support any action by other producers seeking greater amounts than specified or promoted by the OPEC resolutions and that the Gulf states would not respond to any agreement elsewhere with further demands to catch up.

It was this latter point which was the most essential to the companies as they now settled down to negotiate new terms on the short-haul crudes. Their bargaining position was still not overwhelming. The world needed short-haul crude and Libya, in alliance with Algeria, was undoubtedly determined to seek a victory that would set the Gulf in the shadow. But the industry had managed to isolate Libya and Algeria from its Gulf colleagues. A fall in tanker freight rates and an easing in the growth in demand in the West was beginning to reduce the pressure on short-haul supplies and, in both the OPEC meetings and in meetings between the Gulf suppliers of short-haul crude (Saudi Arabia and Iraq through the IPC and Tapline pipelines) and the North African states, some pressure for restraint was exercised by the Gulf countries on their more radical colleagues.

A settlement for the Mediterranean oils, while it did come, took more than three months to complete and was accompanied by an unparalleled series of threats, deadlines and public statements as each individual country in the Mediterranean negotiated its own particular terms.

Libya was still insistent that negotiations could only take place on an individual company basis and that it had the right to a further 5 per cent increase in tax rates as well as retro-active payments for the undervaluation of Libyan oil since the closure of the Suez Canal. Algeria was pursuing its own increasingly tense conflict with the French oil companies over the revision of the 1965 Franco–Algerian pact and in February added a whole new dimension to the oil scene by unilaterally nationalizing 51 per cent of the French oil industry's assets in the country. Iraq reserved its rights to seek back payments on its outstanding claims on the oil industry and made it clear that its approach to Mediterranean talks was going to be very different in spirit to its part in the Gulf negotiations.

On the price question, it was left to Libya to make the running. Again Libya used to the full her previous tactics of seeking confrontation, attempting to divide the companies from each other and constantly taking negotiations to the brink in order to gain the utmost compromise from the companies. By 2 April, after several deadlines had passed and talks had climaxed in a last-minute refusal by the Revolutionary Command Council to ratify the agreements negotiated, a settlement was reached. The oil companies held the line on the tax-rate issue and expressed price increases in terms of specific sums to reflect particular freight and low-sulphur advantages to Libyan oil in order to keep the rises theoretically compatible with the Gulf terms. But they paid heavily for it. Libyan posted prices were raised by an immediate 90 cents with the same escalation clauses as in Tehran, while the tax issue was settled by an agreement to stabilize all tax rates at 55 per cent and to replace the previous tax rate rises with surcharges on the posted price equivalent to the extra income that the previous tax rises would have brought.

It was not until mid-summer that the other producing countries finally agreed their disputes with the oil companies. Iraq took exception to the sulphur premium element in the Libyan agreement and negotiations there took two months before an agreement was reached in June, giving Iraq an immediate 80-cent rise in posted prices with some added sweeteners in terms of cash payments, interest-free loans and a commitment for increases in production. Settlements in Nigeria and Saudi Arabia followed on accordingly but the Franco–Algerian dispute only became more intense, with fierce conflicts on both company and government level accompanied by various restrictions on output and threats of legal action against buyers of sequestrated crude, an attempted embargo on Algerian oil and the withdrawal of French technicians from Algeria before a settlement was agreed between CFP and the Algerian state oil company, Sonatrach, at the end of June. The agreement represented considerable compromise by Algeria on the price and compensation terms but gained for her not only the agreement of CFP (and the French state oil company, Elf–ERAP, in an agreement the next month) to its right to take over control of production but also its right, in the terms set for payments per barrel of exports, to set prices.

The Algerian agreement marked both the end of the extra-ordinary chapter in producer government-induced price rises that had begun in Libya the year before. It also indicated the direction towards national control that the future was likely to follow. For the implications of what had happened in the past two years in terms of producer power and the move to a seller's market was clear for all to see. After a decade in which the main effort of the producers had been concentrated on holding posted prices, and hence the level of their own revenues, at a time of falling crude oil values in the market, the oil states of OPEC had suddenly moved to raise prices again not just in line with market conditions but along the lines which their own concerted efforts and dominant power in the market was able to achieve. The Gulf states alone had raised their per-barrel revenue by an average of 35 cents, or nearly 40 per cent, over the period November 1970 to February 1971, with a firm guarantee of a further 25 cents or so to come over the following four years. Libya's per-barrel income over the same period had risen by nearer 90 cents and the whole series of deals, including its ramifications on Venezuela–Indonesia as well as the short- and long-haul producers of OPEC, was worth something like an additional $20,000m. over the five-year period. Most producer countries could expect a doubling of revenues by 1975 compared to 1970, some, like Libya, could expect a tripling despite their determination to cut back on the rate of growth in production of previous years. And the money, for almost the first time, would have to be found in increased product prices in the market.

If the Tehran and Tripoli negotiations of 1970–71 showed the cost in financial terms of the consumer's dependence on oil for his energy and the Middle East's dominance of those supplies, it showed the cost in other ways as well.

For a generation or more the industrial West, and the developing world, had grown used to ever-increasing volumes of imported oil to fuel their economic growth. Now a buyer's market had swung violently to a seller's market in which the producers held all the cards. Not only was the West irrevocably dependent on Middle East sources as a whole, it was also embarrassingly dependent on any one of those major producers, as Libya had shown. Some of the factors behind this sudden reversal in market conditions – the shortage of tanker-carrying capacity and the tightness in refinery capacity – might be expected to ease. But other factors, like the decline of coal output in most industrialized countries and the failure of nuclear power to fulfil its promise, were not.

In the meantime, world energy demand seemed destined to go on increasing at rates of 4–5 per cent for the foreseeable future and it could only be to the Middle East and Africa that the world could look to supply this increment. Nuclear power would take a decade or more before it began to have any real impact on primary energy patterns. Coal production was difficult enough to hold at its 1970 levels without thinking of raising it in Europe at least. The United States, which for so long could be counted on as a separate self-sufficient entity if not an actual supply source to the rest of the world in times of emergency (as in the 1950s), was now rapidly becoming a major importer in its own right. And oil, despite the increases, still remained a good deal cheaper than any of its rivals.

Given this outlook, few could have thought that the 1970–71 experience was likely to be the end of the story. The implications on the structure of the industry were too great. The producing countries, having sensed their power on prices, were bound to go on to consider the question of control of actual production. The consumers, having so long watched on as observers, were equally bound to consider how best they could intervene in energy matters to protect their oil supplies over the future and ensure a move away from dependence on imports. The oil companies, having found themselves virtually powerless before the producers, were bound to think on their role over a future in which their traditional dominance over oil sources was likely to diminish.

But in looking to the future, the Tehran and Tripoli agreements at least gave the promise that changes could take place within a settled framework for the next few years, that there was time to approach the structural changes in the industry and adapt to the new era. The producers had gained a sharp victory on a scale that would have been unimaginable a year previously. But they had promised stability of supply and responsibility over prices for the next four years and, for that, the price seemed worth paying.

CHAPTER 20

Naked to the conference chamber

'Having faced-down a serious effort to achieve industry
solidarity at the beginning of 1971, the producing
states recognized that industry had no other card left
to play. It was merely a question of time before this
recognition was acted upon.'

*Henry Schuler of Hunt International Petroleum**

It is easy enough in the light of what has happened since to point
out the inherent weaknesses of the Tehran and Tripoli price
agreements; to criticize the complacency which the settlements
aroused amongst many consumers and to charge the oil companies
with short-sightedness for the way in which they obsessively
fought to preserve certain points of principle and precedent that
have come to look less and less relevant with time.

With the benefit of hindsight it is certainly possible to argue,
as some have since, that the industry would have been much
better to have stood up to the producers' threats, to have allowed
them to have taken over production in the Middle East and to
have moved the crude oil trade to a more purely commercial
basis of companies buying from the oil states, thus avoiding all the
problems of participation and nationalization that were to come.
It is equally possible to argue that the consumers themselves
should have chosen this as the time to have faced confrontation
with the producers on the price question rather than – as some
have since suggested – supporting and 'conniving' with the oil
companies in accepting such large price claims.

* Statement before the Senate Foreign Relations Subcommittee on multi-
national companies, 1974.

But such a debate – a debate that is likely to go on for some time both within and without oil companies – ignores the simple facts of political and commercial life as they were at the time. Whatever the broader view, the 1960s had left both consumers and the oil companies singularly ill-equipped to treat the question on a broad strategic scale.

The underlying fact was that both the producers and the consumers had grown far too dependent on Middle East and African production to endanger their growth and economic health for the sake of price demands which they could ultimately afford. Some companies like Shell and BP certainly understood the wider implications of what was happening and were prepared for a showdown in Libya at least. But the primary worry for most companies was their financial position. The changed market conditions during 1970–71 offered the industry its first real chance for some years to turn the tide of declining profitability. A substantial cost rise at the producing end imposed by the oil states could be fairly easily passed on, while the alternative of the threatened stoppage in production or nationalization held much more incalculable risks. The fact, too, that the demands came from the united forces of OPEC rather than individual countries meant that there was no competitive disadvantage for companies that accepted them. Indeed, Libya's demands for even greater amounts meant an actual reduction in the competitive advantages that the independents had previously held.

If the consuming governments had encouraged the oil companies to make a stand or had told them that they would not be allowed to pass on the additional costs, then there may well have been a very different attitude on the part of the oil industry. But the consuming governments, if they were clear on anything, were clear that they were not prepared to support any confrontation with the producers which might endanger their oil supplies and hence their economic growth. And this attitude – apart from the inexperience of the Europeans and Japanese in handling such a crisis – was held for similar reasons to the oil companies'. Oil was of paramount importance to their economies but the cost of it was of much less obvious urgency.

For most manufacturing industries in the West, fuel accounted for only around 5 per cent or less of total costs. Inflation was beginning to push up rapidly the cost of almost every other material so that the rise in oil prices after Tehran and Tripoli, which still did not take the cost of oil to the consumer above what it had cost in 1960, seemed less dramatic than it might have been and certainly not enough to make it a major cost consideration. On broader economic criteria, too, energy as a whole accounted for less than 10 per cent of the Gross National Product in most countries, with oil accounting for only about 3 per cent. Although the impact of the rise in oil prices on import costs was obviously more direct, oil was still no more than about 8 per cent of total imports for most countries (except Japan) and Western Europe and Japan were both broadly in balance in their terms of trade with the producing countries as a whole. When it came to the point of risking supplies for prices – as it did during the haggle over the final few cents during the Tripoli price negotiations in March 1971 – the consumers inevitably responded to the oil companies' request for advice with a general reply that they were not prepared to go to the wall over the issue.

Nor, at the time, did the expectations that Tehran and Tripoli would be followed by a period of stable change seem that naïve. Aside from Libya, which presented a constant threat to any sense of peace and security, most of the producers had adopted a statesmanlike approach to the price question. There was a certain sympathy for their right to share in oil's obvious value in the market and an understanding of their argument that, while the price of goods that they imported from the West was rising rapidly, the price of oil was not. From early on in the negotiations, the producers were careful to make clear that their threats were directed against the oil companies, not against supplies to the consumers, and the promises of future stability that they made at Tehran were backed by the word of the Shah and other leaders.

Just as important, the supply–demand conditions began to ease substantially almost at the same time as the Tripoli price agreement was reached. Demand, after a year of unexpectedly high growth in 1970, began to slow down and the year as a whole recorded one of its lowest growth rates in terms of European and

Japanese consumption for a decade. At the same time, the record freight rates which had done so much to pressure the situation over the previous winter, now collapsed in response to both lower demand and the completion of a substantial *tranche* of new mammoth tankers. Spot market prices for oils, and particularly for fuel oil, which had rocketed over the winter, now fell back by 30 per cent and more during the summer and, in the case of fuel oil, returned to nearly their pre-crisis levels despite the rise in producer government 'take'.

OPEC's next move, to demand participation in the producing concessions during the summer, thus took place in a very different atmosphere from their price demands six months before. Propagated during the 1960s by the cultured and formidable Saudi Arabian Minister of Oil, Sheik Ahmed Zaki Yamani, the ideal of gradually increasing participation by the producing states in the concessions of the majors had been adopted by an OPEC meeting in 1968. It was now revived by the producers as the means by which they would build on their victories at Tehran and Tripoli in terms of actual structural change in the industry at an OPEC meeting in Vienna during July of 1971.

The demand was worrying enough for the oil industry and produced sharp reactions from some companies. Despite the decline in crude oil prices through the 1960s and the increasing tax-take by the producers, the production end of the business remained the profit centre of the oil industry while the extensive concession agreements in the Middle East remained the basic foundation on which the international majors had built up their whole marketing and distribution structure in the Eastern Hemisphere.

No major company, least of all those which had strong crude oil positions in the oil trade, was likely to give up this central area of their business without a struggle. Their fears that the producing countries would not only demand their pound of flesh but also insist that they get it at bargain prices – 'nationalization on the cheap', as one oil company described it – were only made more intense when it was learned that OPEC was considering compensation on the basis of the 'net book value' of the assets rather than taking into account any compensation for loss of

future profit from production. Their suspicion that the whole deal
might simply end in higher prices and greater government take
despite the Tehran and Tripoli agreements was only aggravated
when OPEC started to talk of higher posted prices to compensate
for any fall in the purchasing power of the dollar, which was then
undergoing a serious crisis on the exchanges.

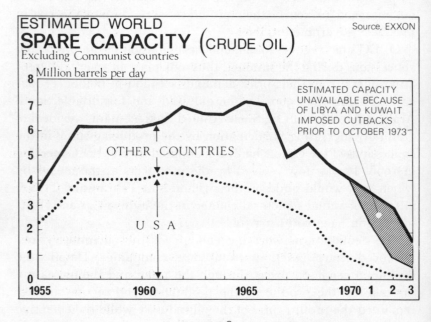

FIG. 8

But many companies at the same time recognized that a
development towards greater control of production by the oil
states was inevitable. The interests of the consumers, primarily
concerned about security of supply, was obviously not as close to
those of the oil companies, arguing about profitability, on this
issue as it had been during the Tehran and Tripoli price talks,
and from the beginning it became clear that the consuming
governments would not support the industry in any dramatic
confrontation over terms.

In many ways, too, participation had strong attractions as a
means of supplying the producing countries with the power they

sought while still preserving the position of the oil companies and the stability of supply. The really major concern of consumers and oil companies alike was that the sudden shift in ownership of large volumes of crude from the companies to the state should not suddenly disrupt the world's oil supply systems and undermine planning for the future.

On this score, participation appeared to offer the same kind of stability for structural changes in the industry as the Tehran and Tripoli price agreements offered on price questions. The increase in producer government share of output starting at 20–25 per cent, would be a gradual one towards an eventual 51 per cent control in the 1980s. Buy-back arrangements could be negotiated to ensure that a portion of the state's new share of oil, at least in the early stages, would be sold back to the companies to prevent too sudden a switch in supply arrangements. Added to this, Sheik Yamani made it clear that it was not Saudi Arabia's intention either to disrupt oil flows to the West or to alter radically existing supply lines by taking on too large a share in the actual sales of oil and investment in downstream operations like refining and marketing, not least because of the fear that the producers might end up competing between themselves and lowering price levels. And in this his views seemed to be further supported by the return of the market to a more buoyant supply–demand relationship, while, on the political side, Iran took the unusual step of publicly dissociating herself from the theme of participation, preferring to go her own route of a contractual relationship that seemed even more moderate in spirit and intention.

If the participation talks thus began in a spirit of relative optimism that a system of stable change could be created, however, it became apparent from early on that there were strong forces working against an easy settlement. Not only were there differences between the attitudes of the consuming governments and the companies on the question, there were also quite striking differences between the companies themselves.

Financial terms that might be acceptable to the Aramco consortium of United States majors in Saudi Arabia, where production was expected to more than double within the space of a few years, seemed very different to companies operating in other countries such as Kuwait, where the prospects of production growth were non-existent. There were differences between the United States and European countries on the principles of participation and legal change in concession contracts and, although the oil companies continued to co-operate in the participation talks through the continued activities of the London Policy group, the atmosphere was by no means as team-spirited as it had been on the price talks.

The producers too were far from wholly enthusiastic about the concept of participation. It had been bitterly criticized by the more radical Arab commentators when first suggested by Sheik Yamani in the 1960s and there were many who still believed that it was too little, too late. Iran's decision to follow a different route raised all the old fears that one country might get better terms than the rest and thus arouse a cycle of competing demands. The short-haul producers like Libya and Iraq clearly wanted more than the initial 25 per cent share sought by the Arab Gulf states. The prospects for complete or semi-nationalization inspired by Algeria remained ever-present and, on top of all this, the demand for higher posted prices in answer to the effective devaluation of the dollar on world markets posed the ultimate question of how long any 'moderate' agreement would last in the face of changing circumstances.

The unfavourable market conditions acted as some restraint on the producers but the year following the Tripoli and Tehran agreements was far from easy. Friction between the companies and the government on questions of exploration activity and investment continued in Libya, while relations between the companies and the government in Iraq were actually worsened by market conditions which induced a fall in shipments of high-price Iraq oil.

In December 1971 Libya shook the oil world with a sudden decision to nationalize BP's Sarir concession in retaliation for Britain's support of Iran's recent seizure of the Tumb Islands in the Gulf. While BP was partly successful in legally blocking attempts by Libya to sell the sequestered oil on the open market, the incident nevertheless showed how little a company could do to secure its rights under such circumstances, how divided the consuming countries were when approached by the British for support and, not least, how easily oil could be used as a political weapon of revenge or pressure on the West.

A month later, after some bitter public argument and a move by Libya to take unilateral action on prices, the oil companies settled with OPEC a new price formula to tie dollar-quoted oil prices to the movements of the currency market. Although the settlement reached was reasonable enough, it inevitably raised doubts about the willingness of the producer countries to stand by any agreement if time and changing circumstances made it look less attractive than it had been when first signed.

In Iraq, relations between the Iraq Petroleum Consortium and the government sunk to an all-time low over the issue of liftings from the East Mediterranean ports (which were well down on the previous year because of changed market conditions and the high cost of short-haul crude oil after the Tripoli agreements) and talks on this issue were then combined with negotiations on all outstanding disputes between the two sides.

Meanwhile the participation negotiations finally got underway in an atmosphere of some tension as the companies at first refused to concede the principle. King Feisal himself was brought in to announce public support of his oil minister and the whole question then relapsed into the old cycle of OPEC threats followed by a move on the part of the companies followed by further threats and the issuing of deadlines for agreement.

Once again the industry found itself in a situation where competing demands by the producers threatened to upset any hope of an overall settlement. Talks in Iraq became bogged down on the question of compensation and the inability of the oil companies to concede any principle in Iraq that might undermine the more general negotiations on participation with Saudi Arabia and the other Gulf States. Iran held off signing a formal agreement on its new terms for fear that it would be outdone by the Arab states. Relations between the Libyan Government and the companies remained strained at the best of times and participation discussions were proving long and arduous.

In June all the old worries about producer–company relations collapsing into individual actions by the producers were raised by Iraq's sensational decision to nationalize outright the Iraq Petroleum Company after more than a decade of continuous and unresolved dispute. Although this was quickly followed by moves to put possible legal action by the companies on ice while talks on compensation were pursued, Iraq's lead was widely acclaimed elsewhere in the Arab world whilst its success in immediately establishing a special oil relationship with France to ensure a market for the seized oil showed just how little support the oil companies might gain from their own governments on such issues.

The turn of the year 1972–73 did see agreements reached on most of the outstanding issues between the oil companies and producer governments but they were uncomfortable ones. A final participation agreement reached in December between Saudi Arabia and Aramco, including some last-minute concessions on timing and price by the companies, seemed to satisfy most of the smaller Gulf producing states, but its attractiveness to the short-haul producers was much less sure and it almost immediately led to a demand by the Shah of Iran for a complete renegotiation of his terms, eventually agreed in the spring. An equally 'final' settlement on disputed questions was reached with Iraq in March, settling compensation for the IPC take-over and agreeing Iraq's rights over the sequestered concession area of 1961. But the participation question for the non-nationalized southern fields remained in abeyance. In April a new adjustment to price was agreed with OPEC after some hard bargaining following the formal devaluation of the dollar against gold. But the talks showed all too clearly that the more radical members of the producing countries – Libya and Iraq in particular – were really after a more fundamental renegotiation of the whole basis of the Tehran and Tripoli agreements.

And behind these negotiations was a much more unsettling and potentially dangerous swing in market conditions back to tight supplies, rising prices at the margin and increasing freight rates, aided by a very substantial demand by the United States for oil imports from the Middle East. On this occasion it was not so much a shortage of carrying or refining facilities that was upsetting trade but a much more basic tightness in oil producing capacity. For almost the first time since the War, the surplus of production capacity against demand, which had been as high as 4–6m. barrels per day or 15–20 per cent of Free World Demand, in the 1960s, was now down to 5 per cent or less.*

Barely was the ink dry on the various agreements, therefore, when market conditions were once again at work to undermine them. The participation settlements gave the producing countries a relatively small proportion of crude to sell on their own account and, during the summer of 1973, these were offered either by auction or on a more informal basis. The mere fact that a new source of oil supply had appeared at a time when supply conditions were unsettled was enough to produce high prices from buyers both from the United States and Japan. In open market sales organized by both Abu Dhabi and Saudi Arabia extremely high prices were fetched. For the first time prices on the open market began to approach posted price levels and this in turn presented serious implications for the whole price and tax-take structure in the Middle East, traditionally organized on the basis that market prices were considerably lower than postings. Abu Dhabi began to press for higher prices on the basis of its low-sulphur oil, which had now become under-priced compared to other oils because of the Japanese demand for non-polluting fuels. By the end of the summer its demands had burgeoned into more general demands by the whole of the producing world for higher prices and a renegotiation of the Tripoli and Tehran agreements.

On the field of participation and government control of output, Libya took the opportunity to pursue its own course with a demand for 51 per cent participation in all concessions. Despite the co-ordinating efforts of the London Policy Group and the 'safety net' clauses protecting the independents by offers of alternative supplies from the majors, Libya once again managed to divide the companies and gain acceptance of her terms from the smaller companies in August and September. The Gulf

---

* This was partly because of the political restrictions placed on output in Libya, where average production was reduced to around 2·2m. barrels per day in 1972 and 1973 against a peak of 3·6m. barrels per day before the June 1970 cut-backs and an expected 4·1m. barrels per day in 1973, and in Kuwait, where the government enforced a stabilization of production at an average 3m. barrels per day in early 1972 against a peak at that time of 3·8m. barrels per day.

countries too began to have second thoughts about the participation deal, which was ratified only by Saudi Arabia and some of the
smaller Gulf producers, and not by Kuwait, who was by the
summer asking for a renegotiation of the terms.

It was against a background of widespread doubts about
whether the long-term price or participation agreements would
survive at all that the Yom Kippur Middle East War broke out in
October and threw oil as well as politics into total confusion,
ending all hopes of a negotiated agreement on these questions and
finally taking the development of producer country power on to
an entirely new plane.

How far the question of oil was mixed up in the Egyptian and
Syrian decision to take the military initiative against Israel may
be open to doubt. Certainly the growing *rapprochement* between
Egypt and Saudi Arabia was a major factor, and President
Sadat secretly consulted with King Feisal before finally taking
the decision. But King Feisal had long taken the view that to turn
off the oil taps in the cause of Arab victory over Israel could prove
counter-productive and it was likely that his view at the time of
this consultation was still that the oil weapon was best used as a
threat to pressure Western diplomacy rather than an accomplished
fact. Over the preceding year he had constantly warned the
United States that her support of Israel was bound to bring
action on oil but he interpreted that action more in terms of a
refusal to increase production rather than actually ceasing it.

Even when the Arab oil producers met under the aegis of the
once-conservative Organization of Arab Petroleum Exporting
Countries (OAPEC) in Kuwait following the war, on 17 October,
there were wide differences in approach to the potential use of
the oil weapon to aid the political cause. The United States'
decision to undertake a massive re-armament programme for the
Israelis undoubtedly induced a general mood that action should
be taken. But, to the surprise of some, it was not the radical
countries that predominated.

Iraq at the outset declared that she was against the idea of any
oil cut-backs, since this would hurt friends rather than enemies,
and preferred instead her own course of nationalizing United
States' assets – as she did with the United States' share of the

Basrah Petroleum Consortium in southern Iraq. Libya's voice appears to have been unusually quiet and it was probably Kuwait, whose National Assembly had long been the scene of some of the most violent anti-Western talk, that took the most radical line. The eventual decision of the producers was thus a mixture of immediate action and the promise of further pressure over the future. Oil production was to be cut back by an immediate 5 per cent with a further 5 per cent reduction to be imposed each month until a full settlement was reached with Israel on the lines of the United Nations' resolution 242. There also seems to have been a general accord that selective cut-backs against specific targets such as the United States might be introduced.

Coincidentally with this and of much greater importance, a meeting of the Gulf states of OPEC, including Iran, also decided to tear up the Tehran price agreements and impose unilaterally new price levels based on so-called 'market prices' starting with a rise of some 70–100 per cent in posting depending on gravity and sulphur differences. The decision had followed a meeting in Geneva between the oil companies and the producers earlier that month in which the oil industry, faced with demands of rises as high as 130 per cent and more, had simply declared that they were unable to negotiate a new agreement of such dramatic impact on the economies and balance of payments of the consuming countries. Although the producers argued then, as they have done since, that the two issues of price and politically-inspired supply restrictions were not connected in intention, the brutal fact was that they had become so in practice.

Within the space of a few days, the producers had effectively wiped out three years of constant and hard-fought effort to arrange a moderate framework within which their new-found power could be expressed. They had finally taken on the right to set unilaterally posted prices; they had firmly thrown oil into the centre of political conflict with the West and they had taken on the right to control destinations for their oil. Once released in an atmosphere of near-exultation over Arab political unity and producer-government influence on the international stage, these developments were hardly likely to be kept within restrained control for all the talk of moderation and justified policies.

And so it proved. Over the next few weeks the situation de-
generated at an unprecedented pace. An embargo on Rotterdam,
the artery of north-west Europe's oil trade, quickly followed a
general embargo on the United States and Western Hemisphere
refineries serving the United States. The first 5 per cent cut was
merged with the next 5 per cent output reduction. Saudi Arabia
and Kuwait interpreted their embargoes as stopping production
of oil intended for Rotterdam and the United States altogether.

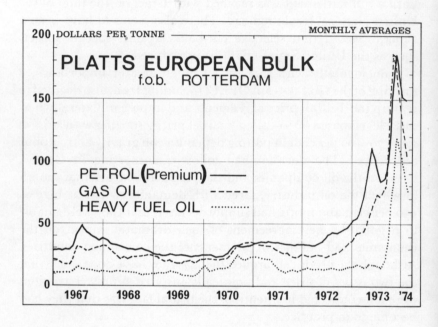

FIG. 9

When the producers next met in Kuwait to consider their position
in early November, it was in a mood intensified by the apparent
unwillingness of the United States to tone down its support of
Israel and the lack of real response by Europe to the first cuts.
With the Kuwaiti and Saudi Arabian moves already involving
effective reductions in output of 25 per cent or more, it was agreed
to level out all restrictions at 25 per cent with the promise of
more to come. At the same time, following the lines already set
by Saudi Arabia, other producers began to adopt systems of

categorizing consumer countries and treating their oil supplies with differing degrees of restriction.

This in turn threw the world's oil trade and the consuming countries into a welter of confused responses and nervous deliberation. The major oil companies themselves attempted to share out their oil supplies as best they could in the face of differing instructions from the individual producers and grudging consent from individual consuming governments, some of whom had been told that they would get 'favoured treatment' from the producers. Attempts to gain sharing agreements between the consumers through the OECD or EEC failed before consumer fears that any concerted action on their part would only arouse more extreme action on the part of the Arabs. Individual countries took action to control product exports, to restrict home demand by allocation, rationing or bans on Sunday driving – many of them for the first time – and several made early trips to the Middle East in order to seek special supply arrangements with the producers.

Meanwhile, the marginal markets for oil went into a wild spiral of escalating prices made only the worse by the failure of the consuming governments to ensure any stable supply or price arrangements amongst themselves. The Rotterdam and Italian export markets saw prices rocket to unprecedented levels as brokers held on to shipments waiting for the market to hit its peak, consumers busily built up stocks against the coming disaster and new supplies of marginal oil began to dry up altogether. In the crude oil market, a series of auctions held by the producing countries brought forward some truly staggering bids. Tunisia gained over $12.60 for a relatively small sale – nearly double its posted price. A month later Nigeria had gained an even more astonishing bid of $22 per barrel for its crude oil – two and a half times its posted price – and then, in the most astonishing sale of all, Iran gained bids of up to $17 per barrel and more for a much more substantial sale of its state-owned oil, no less than three times its posted price.

By early December there were signs of some softening in the attitude of the Arab producers at least on the cut-back question. Having originally felt that the consumer response had not been strong enough, a visit to Europe by Sheik Yamani and Belaid Abdessalem of Algeria convinced them that they might now be going too far and that the exercise was beginning to prove counter-productive in the sense that all that could be done against the United States had been done by the embargo while Europe could really do very little at all to help the Arab cause.

But prices were a different question. Under OPEC's stated aim of tying posted prices at around 40 per cent above market prices in order to re-establish what they claimed to be the old 1 to 1·4 ratio between market and postings, prices for 1974 should have been set at around $25 per barrel compared to $1.60 only three years before. This was obviously out of the question but when OPEC met to discuss the situation in December it was with mixed feelings. Some, like Saudi Arabia, felt that the producers should not take too much of a commercial advantage from artificial conditions of shortages induced primarily for political purposes. Others, like Iran, Iraq and Algeria, with greater need for revenues, felt that a dramatic rise was in order and it was under the leadership once again of the Shah that a final decision was made to double prices again on the simple basis of ensuring a producer government take equivalent to the cost of producing oil from shale. A consuming world, already profoundly perturbed by the political implications of oil restrictions, thus woke up to an oil price rise far beyond its worst expectations – one that involved a tripling of landed oil prices within a few months and one that would entail virtually every importing country going into a serious balance of payments deficit.

December 1973, however, proved the height of consumer discontent. As in the crisis of 1970–71, the seasonal pattern of oil trading, with its peak demands at winter, combined with the volatility and disproportionate influence of the marginal oil markets, tended to exaggerate the degree of real pain and swing the pendulum of expectation too dramatically in one direction or another. The short-fall in supply was real enough, at one stage reaching some 6m. barrels a day or some 17 per cent of the expected world's oil trade at the time. But this was compared with previous expectations of a cold winter and surging demand. In fact neither happened. The winter, as in the previous three years, proved unusually mild. The action of the major oil companies was reasonably effective in 'sharing out the misery' evenly between consumers. Government action to restrict demand, coupled with a substantial reduction in demand due to the size of the price hikes, served to suppress consumption to near the level of supply, while very ample stocks in many countries provided enough oil to make up the difference.

Cost break-down of an average barrel of oil sold in Western Europe, 1973–74 (in $ per barrel)

	July 1973	January 1974
Producer government 'take' (including 25 per cent participation)	2·30	8·48
Production cost	0·19	0·18
Transport	1·00	1·50
Refining	0·53	0·55
Distribution	2·20	2·20
Consumer government taxes	7·45	7·70
Oil company margin	0·58	0·88
Average recovered price in the market	$14·25	$21·49

As the spring of 1974 saw the end of winter, so did the worst of consumer fears about the oil situation dissipate themselves. At the turn of the year, the producers had decided to ease back the cuts to a level of 15 per cent below September with a wide range of exemptions to particular countries. By April the embargo on the United States was lifted at Saudi Arabia's insistence, production rose and the were over in all but name as individual producers crept above their self-imposed limits and consumer demand started its rapid fall to the seasonal summer low.

Production, even allowing for little or no growth in the output of some Arab countries, was still more than adequate to cope with demand as now expected. The oil crisis had acted on a situation in which world trade and economic growth in the West had in any case seemed destined to falter. Output from countries not involved in the cuts, like Iran, Nigeria and Iraq, was still growing at rapid rates. Price itself had become a major restraining influence on consumption. Meanwhile, on the marginal market, the pendulum was swinging back again. Brokers with oil supplies now needed to sell before the summer came upon them but there were few takers at the high prices, partly because many consumers had built up high stocks themselves. The spot market for products in Europe began to fall almost as dramatically as it had risen. This in turn encouraged buyers of crude oil both to walk away from some of the contracts paid at top prices and to offer less for new auctioned supplies. Kuwait and others found it impossible to sell all the oil at the prices that they demanded, let alone the prices they had briefly gained at the height of the crisis. Once again, just as in 1971, the situation seemed destined to settle down to more normal trading conditions.

And yet, what were more normal conditions? As the dust began to settle in the aftermath of the 'crisis', the consumer might feel more relaxed but for governments and industry alike there were nothing but questions hovering over the oil trade – about prices, about supplies for the future, about the structure of the industry and the role of the companies, about the motivations of the producers and about the place of oil in the world's energy patterns.

The producers, having experienced such power in the market, were moving rapidly to take full effective control of oil production through revised participation agreements, outright nationalization and entirely new contract terms. Having once unsheathed the sword of economic warfare, no one could be quite certain when it might be used again. Having taken control of prices, no one was quite sure how they would use this power over the future, how long their unity in OPEC would last what they would do with the enormous surplus funds that some would gain from the increased revenues and how they would sort out their relations with the industrialized West and, more painfully, the rest of the developing world.

The consumers, meeting together for almost the first time at top international level in the Washington conference on energy in February, seemed equally uncertain as to their response to the situation. While some, led by the United States, urged a grand alliance between consumers to force down prices and guarantee co-ordinated restraint in supply conditions, others felt that direct bi-lateral deals with the producers were the only sensible route,

11. Disaster: (1) *Torrey Canyon* breaks up

while still others saw the advantages of particular initiatives from Europe rather than the United States. While some were acting to restrain demand, others felt that prices alone would be the effective counter-weight to the growth in oil needs. While some felt that prices of oil were too high to sustain over the longer term, others thought not. While some felt that the aims of the producer and consumer were contradictory and could never be fully reconciled, others argued that co-operation between the two could still provide a secure framework of supplies and fiscal stability for the future.

And in between the two were the oil companies, enjoying the first really substantial surge in profitability in over a decade, but everywhere under pressure – uncertain of whose oil they were actually selling now that the producers were demanding back-dated control, confused about prices and their future direction and abused by both producer and consumer alike for their role as the middleman of the crisis. The last effort at reasoned price negotiations with the producers at Geneva at the beginning of October 1973 had seen them bow out of this, their traditional role, and admit that they were no longer able to act as a counter-vailing force to producer demands. The last effort at sharing out oil supplies during the winter of 1973–74 has seen them reluctantly admit that they no longer had the power or the resources to do this job effectively without the complete and direct backing of the consuming governments. An industry which for fifty years had argued that government intervention would only serve to damage the interests of the consumer was now saying that it could not manage the job effectively without that intervention. After three years of constant change and continuous bargaining under threat and demand, the seventh veil had been removed to reveal an industry shorn of its traditional power as the buffer between producer and consumer and bereft of the trappings of supra-national influence and oligopolistic control of the market which had, rightly or wrongly, been so long associated with it.

CHAPTER 21

The energy crisis

'This Time The Wolf Is Here.'

*James E. Akins**

The most fundamental question of all to be raised by the dramatic events of 1973–74 is perhaps simply whether the drama has reflected a genuine and deep-seated energy crisis facing the world or not. For some, certainly, the problems of oil in the early 1970s have proved yet one more example of the way in which a greedy and wasteful consumer society has been using up the world's natural resources to a point where these resources are no longer sustaining the growth in consumption and where the world is in real danger of being unable to get the energy it needs to continue its current way of life. For others, the difficulties of the 1970s have been the product less of a fundamental shortage of resources than a temporary and exaggerated imbalance between supply and demand caused by bad consumer government policies, a cartel of producers and the self-interested hysteria of the oil companies. For some, price alone will be quite sufficient to bring supply and demand into balance, for others it is government intervention that will have to provide the answer while for others still the only thing that will avert disaster will be a revolutionary change in the structure of the consumer society.

There is ample evidence for all these views, and indeed they may not be as incompatible as they seem on the surface. Certainly, the constant flow of cheap oil on to the world's markets since the Second World War has encouraged a 'profligate' use of energy

* The title of a prophetic and highly percipient article by James Akins, then Director of the Office of Fuels and Energy, United States Department of State, in *Foreign Affairs*, April 1973.

in general and oil in particular whose consequences are only now being brought home to the consumer. To use the well-worn example of the lily in the pond that doubled its size every week: no one worried so long as it was small. But one week it grew to cover a quarter of the pond. The next week it covered half the pond and then people realized that they only had another week before it covered all the pond. The same is true of energy consumption. As the scale of use grows so does the volume implications

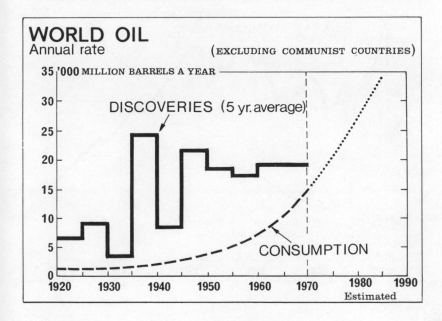

FIG. 10

of a growth rate in energy of 5 per cent every year, involving a doubling of consumption every fifteen years, and a growth rate in oil consumption of 6–8 per cent, involving a doubling of use every ten years. Put another way, the world on historic growth rates will use as much oil over the next ten to fifteen years as it has in the last hundred years. The resulting strains have been amply illustrated by the constant crises of the early 1970s.

But even accepting this argument, it would still be hard to assert that there is any fundamental shortage of energy resources as such or even of fossil fuels (coal, gas and oil). Worldwide reserves of coal alone are estimated to be about 4–8 trillion (million million) metric tons, enough to support world consumption at present rates for a thousand years, or, put more realistically, sufficient to allow a rapid expansion of coal consumption by some sixfold over the next 150–200 years before the actual amount of coal in the ground restrains growth.

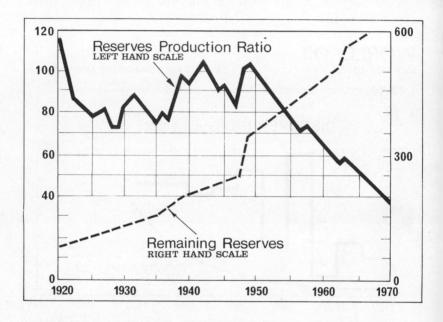

FIG. 11

Oil reserves worldwide can equally be expanded several times over by the addition of the immense quantities of unconventional sources for oil such as tar sands, oil shale and heavy oil deposits. The great expanse of Athabasca tar sands in Alberta, Canada, for example, is estimated to contain some 635,000m. barrels of oil, or about the same quantity of proven conventional oil reserves around the world. Oil shale deposits in Colorado in the United States are thought to contain around 1 trillion barrels of oil, or double the total of proven conventional oil. Overall, including the Orinoco Tar Belt in Venezuela, the Brazilian oil shale and other similar deposits, world shale oil reserves could amount to some 5 trillion barrels of oil.

These are, of course, figures for ultimate reserves rather than reserves which might be produced under current technological and economic conditions. But even on the basis of today's technology, estimates suggest that synthetic crude oil recoverable from tar sands outside the Communist world might amount to some 250,000m. barrels, rising to double that as the technology improves over the future, while some 50,000m. barrels might be

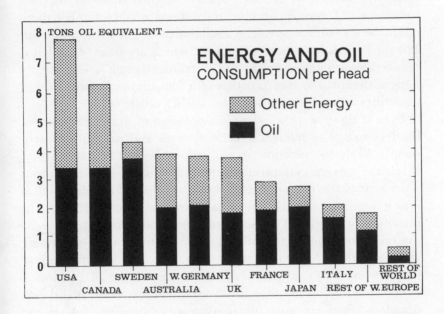

FIG. 12

recoverable from shales under current techniques. Adding all these estimates of fossil fuels together, including conventional crude oil and natural gas reserves as well as coal, shale and tar sand reserves recoverable at reasonably high rates of technology, a grand total of something like 5 trillion tons of oil equivalent can be reached. Set against a consumption rate at the end of this century calculated on the basis of a 5 per cent annual growth, this would be enough to last the world some two hundred years and the figures could easily be improved if breakthroughs in recovery techniques are made, a lower rate of growth in consumption is assumed or if more detailed surveying changes the picture on reserves of any one fuel.

Nor is it likely that the world will be calling on fossil fuels to this extent by the end of the century. Nuclear power, despite all the problems and delays in getting it underway, should begin to expand rapidly in the 1980s and, for many, will provide an almost boundless supply of reasonably-priced energy over the long-term future. It is not without its own demand for fuel (uranium), of course, and on the official reserve figures there might be a problem over the availability of this fissile material. But the figures for uranium reserves, which are based on what is recoverable at current prices and in 'politically safe' areas, greatly underestimates the true position of a fuel that is found in small quantities at least throughout the world's crust.

Even if there were a long-term problem of uranium reserves, the technology of nuclear reactor systems and uranium enrichment is likely to overcome it well before the end of the century. While the current generation of thermal reactors uses only about 2 per cent of the nuclei in its fuel, a great deal of development is now going into construction of 'fast breeder' reactors such as the prototype at Dounreay in Scotland which can 'breed' fresh fissile material from the 'fertile' nuclei in uranium in a self-sustaining chain reaction that could potentially increase the efficiency of fuel conversion to 50 per cent. With these types of reactors – which could begin to come into widespread use later in the next decade – there are more than enough uranium reserves in the world to last several hundred years. And farther in the future than this – although still facing formidable problems of design and engineering – there is now increasing research and considerable optimism about the development of nuclear 'fusion' which could provide unlimited energy for a million years or more converting ocean deuterium.

While nuclear power is generally held out to be the main hope for the provision of bulk power over the future, other 'unconventional' systems such as tidal, geothermal and solar energy are also attracting substantial interest as potential long-term sources of power. Although restricted in a sense by geographic limitations, all hold considerable potential. A pilot plant for using tidal power is already in operation in France and, various suggestions have been made about the use of waves or temperature differences in the sea to generate electricity using relatively basic technology.

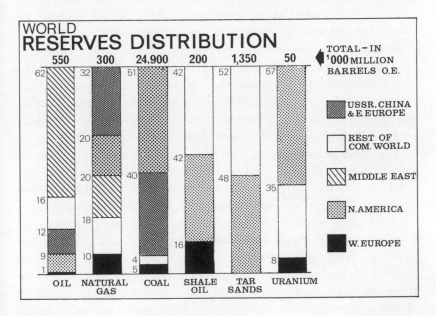

FIG. 13

Even wider prospects are being held out for geothermal energy, which uses the heat of the earth's interior by tapping steam from underground hot water reservoirs to generate electricity and the water itself to heat houses and other applications. Power stations based on this source have long been working in Iceland, the United States, Mexico and Russia and interest is now growing in the development of such reservoirs elsewhere and possibly extending their use by creating artificial reservoirs blasted out underground.

Solar energy, meanwhile, has become for some at least the most attractive of all long-term energy sources, albeit with many doubters as well. Theoretically the scope for development is enormous. The heat from the sun represents the principle input of energy into the earth on a scale many thousand times more than the world's conceivable energy needs in the future. It is already used in a number of small-scale schemes to heat houses, provide hot water and generate electricity. The problem is the collection of heat, which is by nature diffuse and subject to climatic conditions, on a large scale. But the technology for photothermal and photoelectric conversion is developing. A prototype power plant is planned in the United States for operation in 1976–77 and there are now suggestions that the climatic problems could be overcome by collecting solar rays in space, converting them to electricity with solar cells and transmitting power to the earth by micro-waves – a scheme of horrendous cost not totally beyond the bounds of ultimate possibility.

The great attractions of solar energy, especially for conservationists, are that it is virtually pollution-free, that it could provide a solution to the current problems of uneven distribution of fuel resources around the world and that it would encourage decentralization and self-sufficiency, quite apart from its low running costs. Some of the same advantages could also be ascribed to chemical energy systems under consideration for the long-term future such as those based on hydrogen, which could be produced from water using nuclear power, or methanol, which could be synthesized from carbon dioxide and hydrogen with few resource worries. On the other side of the fence, very considerable research is now being undertaken into the development of more efficient energy-using technologies such as battery cars, or fuel cells and magnetohydrodynamics.

But, however exciting these prospects are over the long-term, they are most unlikely to be of much use in the immediate future or to provide a serious short- or medium-term alternative to the world's current dependence on oil as its major source of primary fuels. Coal in many areas of traditional strength, particularly in Europe, has been allowed to decline to a point where it will be difficult enough to keep output at current levels, never mind attempting to turn the industry round to a new era of expansion. There are considerable, although less discussed, doubts about getting a new generation of miners down the pits and the development of less labour-intensive technologies for mining is still surprisingly under-developed. In the United States, where the main hopes for a rapid expansion of coal are concentrated, there are severe environmental objections to large-scale strip-mining, while in Europe, where the coal is in deep seams, there are continuing problems of cost, despite the price rises on oil. Above all, there remains the overriding problem of 'lead-times' and finance in getting new, large-scale coal developments underway in the areas of greatest promise like the United States and Australia.

The difficulties of 'lead-time', of cost and of environmental restrictions are even more apparent when it comes to considering the newer forms of energy such as nuclear power, solar energy and the development of non-conventional oil sources from tar sands and shale. Despite the high hopes that nuclear developments have raised in the West over the last two decades, the actual progress of nuclear programmes has fallen well behind schedule. Britain, which had hoped to be generating some 11,000–12,000 MW of electricity from nuclear power stations by 1974, was in fact generating only about half that. The United States, which had planned to have some 53,000 MW of nuclear capacity installed and working at the same time, had in fact achieved less than 15,000 MW.

And the chances of pulling this situation round within this decade are very slim indeed. Part of the reasons for past delays has, it is true, been due simply to the fall in oil prices during the 1960s, which tended to sap the enthusiasm of governments for the highly expensive nuclear effort. But it has also been due to the failure of governments and power authorities to understand the engineering problems associated with scaling up initial designs and prototypes to full commercial operating size and their under-estimation of the strains that massive programmes would put on the contracting and supply industries.

The same factors still apply to nuclear and other technologies. It takes something like seven years between the planning of a nuclear power station and its actual installation and past experience has shown that it is extremely difficult, if not actually harmful, to attempt to speed this up. Progress has been further delayed, and will continue to be delayed, by serious environmental objections to the siting of nuclear power stations and deeply-felt, and partly justified, worries about safety in the event of a leak.

On 'pre-crisis' estimates, nuclear power, which at present accounts for only some 3 per cent of the world's primary energy demand outside the Communist areas, is unlikely to account for much more than 7–8 per cent by the end of the decade. Even assuming a rapid expansion after that, it will still probably not take up more than 15 per cent of demand by the mid-1980s before making a really strong impact on energy patterns in the 1990s.

Synthetic crude oil production, solar energy, tidal power and the development of more efficient energy systems for transport and heating all suffer from the same difficulties. Plants producing small quantities of oil from the Athabasca tar sands and pilot plants to produce oil from shale in the United States have been working for some years. But there are still formidable problems of designing and constructing large-scale commercial plants to mine or produce *in situ* oil from these sources. Investment is certainly beginning to flow in this direction and in others, but the actual volumes of energy output from these sources or geo-thermal power plants or solar energy will be relatively small in comparison with total needs for a decade or more ahead. And the time could be a great deal longer if the current design and engineering problems are not overcome soon.

Just as in the past twenty years, so for the next decade at least, therefore, it will be oil and gas that will have to 'fill the gap' and sustain much of the burden of the expected growth in demand. Whether they will be able to do so, and whether the major producers will allow the rate of expansion needed, is the question now worrying industry and governments alike as the crises of the last few years have finally brought home to the consumer the possibility that the finite reserves of crude oil will no longer manage the task of meeting demand growth.

Worries about oil reserves running out are not, of course, a new feature of the oil industry. The calculation of oil resources is always a somewhat tenuous business, depending, as it does, on estimates of recoverability that are in themselves economic judgements of what can be done at present prices and present costs and on speculation on the future that is based on past trends and which could be upset at any time by the discovery of new oil provinces such as the North Sea. Shortages of oil were widely predicted in the 1920s, again in the 1930s and again just after the Second World War. No shortages occurred and each time the reserve calculations, far from falling, were dramatically expanded within a short time with new discoveries and the constant re-evaluation of the massive Middle East reserves after the War.

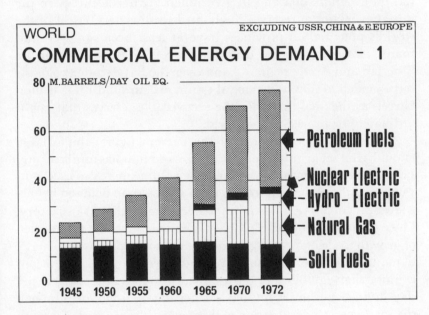

FIG. 14

Where estimates in 1938 put world proven recoverable reserves at 31,000m. barrels, or enough to supply demand at that time for fifteen years, in 1950, reserves were assessed at 95,000m. barrels of recoverable oil, enough to last twenty-five years at prevailing rates of production and by the end of the 1960s, estimates, following further re-evaluation of Middle East potential, suggested world proven reserves of about 529,000m. barrels, or enough to last for thirty-three years at the output levels of the time. Latest estimates, for the end of 1973, put the figures even higher at a worldwide total of 634,700m. barrels of recoverable oil, the equivalent of nearly thirty-five years of output at current rates. More sophisticated oil-finding techniques, the move off-shore and the re-evaluation of existing reserve figures in the light of different economic circumstances and more sophisticated methods of recovery can all be expected to increase substantially the amount of oil found over the coming decade or more.

But the improvements in seismic and exploration technology are also serving to define more and more closely the areas of potential oil accumulation and to give much greater accuracy to the initial estimates of reserves made on first discovery. Most of the world's potential oil provinces have at least been surveyed and many of them actually drilled – save for Russia's immense Siberian and Arctic regions – and over the last decade geologists have tended to stick at a general estimate of around 1·6–2 trillion barrels as the final figure for the recoverable reserves that might ultimately be proved in the world.

At about double or more of the current figure, this is large enough. But set against the scale of demand that has until recently been developing in the world, and the volume increases in oil output that will still be required to meet a growth in demand of only a few per cent per year, the position becomes much less satisfactory. Over the last two decades oil has taken on almost the full burden of growth in primary energy requirements at a time when energy demand itself has been rising by some 5 per cent. This has meant a more than doubling of oil use from around 22m. barrels per day in 1960 to some 53m. barrels per day in 1972, or over half the total energy consumption of the world. If these kind of growth rates were to continue and oil was still to shoulder the main burden of consumption growth, it would mean a fantastic increase of over 30m. barrels per day in oil output during the 1970s to a level of well over 80m. barrels per day in 1980 followed by a further rise of 20m. barrels per day to 100m. barrels per day or more during the following five years, even at a reduced rate of growth. In other words, a new North Slope of Alaska would have to be found every year and a new North Sea every three years just to keep pace with the annual increment in demand.

The most disquieting feature of the oil scene today is that this kind of discovery rate is simply not occurring. Just as with every other energy source, oil is entering a phase where the industry has to run harder and harder to keep still. The prospects of finding another Middle East grow dimmer as the search covers more of the world. Exploration is now moving into areas of the off-shore, in the Arctic and in the jungle where investment is large, the rate of drilling wells is slow because of the physical conditions and the lead-times between discovery and start-up to production are running to five years and more.

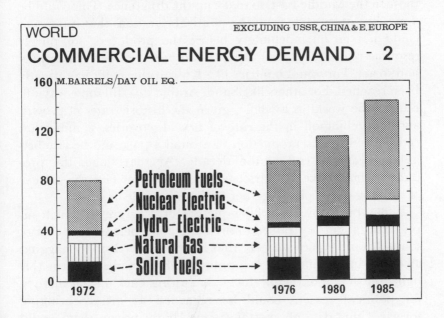

FIG. 15

A more precise analysis of recent trends, backdating reserve estimates of major fields to the date of their discovery, has shown that the period of really substantial increments to world reserve figures, largely through the constant revision of Gulf estimates, in fact reached its climax in the mid-1960s and that the increase in production began to exceed the net addition of new reserves in the world outside Russia and China at the end of the 1960s. Since then the annual rate of new discovery, which has added about 18,000m. barrels of recoverable reserves per year on average since the Second World War, has been dropping and with it the ratio of world reserves to production.

Even assuming that high prices and more sophisticated exploration techniques enable the rate of annual discoveries to be restored to its former level, this would not be enough, at least on pre-crisis estimates of growth, to keep pace with the rise in demand during the 1970s. The world would then have had to depend on an increasing rate of depletion of existing proven reserves, particularly those in the Middle East, to make up the difference. This could be done. But quite aside from the political questions that it poses, it could not go on indefinitely before the producers, for purely resource reasons, felt that a point had been reached where they had to act. For some countries like Kuwait this point has already been reached. For others like Saudi Arabia it is still some way off. But for the world as a whole, given the historic rates of growth and a levelling-off in the rate of new discoveries, a minimum ratio of reserves to production of around 15:1 could be reached as early as the end of the decade. At that stage, the producers would have to restrict further growth or face the danger of running out of oil altogether within a decade or so.

It has been this disturbing picture of a future in which oil resources might be strained to breaking point before other energy sources could come into play that has been increasingly nagging at the industry since the beginning of the decade. And, in this fight, the action of the producers in cutting back output in 1973 for political reasons could be regarded as merely bringing forward the day of shortages and high prices that would have occurred in any case for natural reasons some five or six years later.

Yet in doing so the producers may well have also brought forward the solutions to that crisis. Speculation in terms purely of potential demand and resource availability is useful as a means of pointing out what may come to pass if nothing is done about it. It is less useful as a method of predicting what will actually happen. Both supply and demand are ultimately a function of price as well as physical resources, and this is especially true of fuels, where reserves are estimated in terms of what is economically recoverable at certain prices and future demand and supply patterns are predicted on the basis of extrapolation of past trends, relating oil demand to economic growth and rates of production to price and demand.

What the estimates of energy resources over the past few years have done is to suggest that oil is now reaching its peak as the paramount fuel in the world's energy picture and that its previous growth trends cannot continue without serious strains emerging politically and economically. The rate of growth in demand is now beginning to exceed the rate at which new reserves are being added and, taken to its logical conclusion, this must lead to a point at which demand itself will have to be curbed if actual shortages are not to appear.

While this does much to explain the background of the recent oil crises, however, it ignores the effect that these crises will have on supply–demand patterns. Already, for reasons both of the oil crisis and other economic developments, the Western world has been thrown into a 'no growth' situation for 1974, thus putting off the date of predicted troubles by at least a year. More than that, it has also set in motion some long-term trends towards a lower growth rate in oil consumption by raising prices to a point where the consumer has to think not only of conversion to other fuels but also of reducing his own demand for oil where this is possible.

How effective this reduction in consumption will be is still extremely difficult to predict. But the immediate experience of the aftermath of the crisis suggests that it might be substantial in terms of cutting out the luxury use of oil for heating in homes, offices and factories; in forcing major consumers to think of more efficient ways of using their oil and investing in better equipment; in lowering the consumers' expectations of the personal comfort and mobility that cheap petrol and heating oils have brought in the past and, not least, in encouraging a move towards smaller cars with lower fuel consumption. All these could have a fairly rapid impact on oil consumption rates while the presence of high prices could well induce further investment in fuel-saving schemes such as 'total energy' and in developing more efficient cars, generating equipment and manufacturing processes that would keep up the pace of change towards less wasteful and hence lower rates of oil use over the medium- and long-term future as well.*

On the other side of the coin, high prices and serious concern about security of supply will undoubtedly divert – and have already done so – resources in the consuming areas into the more rapid production of alternative fuels and alternative sources of oil. This will take time of course. But in the short-term there are strong hopes that some additional oil could be produced in the United States, for example, by opening up naval reserves and by producing from small fields previously considered uneconomic. There are hopes that coal production in the United States and in other major coal-producing areas could be stepped up substantially over a period of five to ten years while, within the same sort of timespan, greater output might be achieved from shale and tar sand sources. Meanwhile, the rate of exploration, the level of investment in off-shore development and the amount of money being poured into research and development of nuclear and solar energy, the conversion of coal to oil and gas and other energy technologies, will all undergo dramatic expansion.

Whether this will produce the glut in fuels that some economists foresee following the crisis of 1973–74 is open to debate. There are still formidable obstacles of engineering and construction to the rapid development of alternative energy sources. At the prices reached in 1974 of some $10–12 per barrel delivered in Europe and the East Coast of the United States, oil has already passed the $6–8 mark at which oil-from-coal and from tar sands and shale become highly competitive, and is well beyond the point at which nuclear energy becomes a viable economic alternative. But the sheer practical difficulties of building the size of plant needed in these fields and of organizing the management, supplies and contractual forces necessary to build them at speed are likely to remain a restricting factor for some time to come. On the demand side, too, there must still remain doubts about

---

* Estimates made immediately after the cut-backs of 1973–74 suggested that price alone was causing a drop of 10 per cent in European consumption and that previous estimates of demand in 1985 could be shaved down by as much as 10m. barrels a day, or over 10 per cent, or perhaps even more because of lower growth in oil demand in the West and the more rapid development of alternatives.

the long-term ability of consumers to save on oil and overcome their natural tendency to revert to more wasteful ways once the immediate crisis is past.

Much in the end will depend on how far pressure on prices is kept up and how the rate of price increases on this fuel compares to other commodities. Much will also depend on how far the major exporters, particularly Saudi Arabia, are willing to go on increasing production even at the cost of weakening prices and how far the major consumers like the United States really ensure the diversion of resources necessary to reduce their dependence on oil imports. If the United States really does achieve virtual self-sufficiency within its target of 1980, the Middle East producers do keep up their traditional growth in output, and all the hopes for other energy sources are realized, then a surplus could well occur, at least in the next decade and quite possibly within this one. But if improving supplies coupled with public apathy and high inflation sap the consumer will, and if investment in alternative energy sources does not bring forth the expected results, then the day of reckoning may simply be put off for a time rather than avoided altogether.

CHAPTER 22

Producer power

> 'The raw material producing countries insist on being
> masters in their own houses. . . . The developing
> countries must take over their natural resources, which
> implies, essentially, nationalizing the exploitation of
> these resources and controlling the machinery governing
> the determination of prices.'
>
> *President Boumedienne of Algeria**

As long as oil remains the balancing fuel in the world's energy
economies, it will be the major oil exporting countries of the
Middle East, and, to a lesser extent, Africa, Asia and Latin
America who will hold the ultimate key both to prices and
supplies. The crises of the Tehran and Tripoli price negotiations
and the grand drama of the supply cut-backs and price increases
of 1973 have already shown just how far this power extends in
political as well as economic terms. The question now is in what
way they will use this hard-won control of the world's oil trade over
the future – whether they will produce solely in answer to their
own internal needs or respond to the world's much greater needs
for their oil; whether they will extend their power over production
to greater participation in the other phases of the oil business as
well; whether they will use their new-found wealth to expand
their influence through the developing world or keep most of
their investments in the West and, not least, whether the use of
oil as a political weapon will be used outside the forum of the
Israeli conflict over the future or dropped from the scene for all
time.

The uneven distribution of the world's oil reserves which has

* In a speech to a United Nations General Assembly meeting called to discuss
the problems of raw materials and development in April 1974.

given such extraordinary economic power to such a small and volatile area of the world is not an entirely new feature of the oil business, of course. Before the Second World War it was the United States and the Caribbean (Venezuela in particular) which provided most of the oil going into world trade. In the 1950s it was the Middle East and the Caribbean which predominated, and in the 1960s it has been the Middle East and North Africa which have provided the main international flows of oil.

Growth in producer oil revenues, 1963-72

Country	1963 Revenues ($m.)	1963 Revenues per barrel (cents)	1972 Revenues ($m.)	1972 Revenues per barrel (cents)
Kuwait	557	74·3	1,657	140·9
Saudi Arabia	502	78·7	3,107	143·7
Iran	398	79·7	2,380	135·8
Iraq	325	80·7	575	150·7
Abu Dhabi	6	36·4	551	143·4
Qatar	59	84·2	255	144·5
Others	13	79·3	222	120·6
Total Middle East	1,860	77·7	8,747	140·7
Libya	109	65·1	1,598	196·6
Algeria	—	—	700	187·7
Nigeria	—	—	1,174	187·0
Venezuela	1,106	98·6	1,948	171·9

Source: *Petroleum Press Service*, November 1973

But it has only been in the last decade or so that oil imports have established such a predominant role in the economies of the major industrial areas of the world outside North America and only in the last few years that demand has begun to strain the available production capacity of the producers. In 1925 less than 16 per cent of the primary energy consumed in the world crossed foreign borders. In the mid-1960s it was about one-third and by the end of this decade it may well be two-thirds. Japanese imports alone have more than quadrupled from around 1m. barrels per day in the early 1960s to some 5·8m. barrels per day in 1973. Western European imports meanwhile have nearly tripled from 5m. barrels per day to over 15m. barrels per day and, overall, the amount of oil being moved into international trade has increased from 2m. barrels per day before the Second World War to 20m. barrels per day at the end of the 1950s, or some 45 per cent of total world oil consumption, to over 34m. barrels per day in 1973, or some 60 per cent of world oil consumption, making oil by far the largest item of international commerce.

And the sources for this growing international trade in oil have tended to become more and more concentrated on a small number of producing countries banded together in the Organization of Petroleum Exporting Countries. Of total world oil exports in 1973, no less than 60 per cent, or 20m. barrels per day, came from the Middle East, while a further 10 per cent came each from the Caribbean and North Africa. All the major contributors to this trade have by now become members of OPEC, which as a group now accounts for some 90 per cent of all oil going into world trade, controls virtually all the oil supplies of Europe and Japan and contributes over half the total primary fuel supplies of all types used by these areas.

For the future, there seem few ways of avoiding this growing dependence on OPEC over the medium-term or to prevent the growth in trade being further concentrated on exports from a few central exporting countries in the Middle East. Despite the discoveries in the North Sea and Alaska, the major consuming areas – including the United States for the moment – are having to look mainly to oil imports to meet their growth in demand. Some idea of the potential scale of this trend can be gained from the 'pre-crisis' estimates of the oil import requirements of the industrialized world during the 1980s, by which time United States' requirements alone were expected to grow to 14m. barrels per day by 1980 and 16–17m. barrels per day in 1985, Japanese North Sea production, were expected to rise to 19m. barrels per day in 1980 and 12m. barrels per day in the mid-1980s and western European imports, although being shaved down by North Sea production, were expected to rise to 19m. barrels per day in 1980 and as much as 24m. barrels per day in 1985. High prices and consumer government action, particularly in the

United States, will almost certainly prevent a growth on anything this scale occurring, but it is difficult to foresee a complete release of the West from some kind of growth in import needs, at least during this decade.

Under current conditions only the Middle East, with its vast reserves of oil and its low costs of production, can possibly cope with the demand. Nearly 60 per cent of the world's total proven reserves lie in an area 800 kilometres by 500 kilometres around the Gulf. Total recoverable reserves in the Middle East, at some 350,000–400,000m. barrels, are enough to last fifty-five years or more at current rates of production and the figures could well prove well short of ultimate oil resources in an area where there has been little need until recently to make definitive assessments of the ultimately recoverable potential.

Compared to these figures, most other areas pale into insignificance. Africa, although it has some 15 per cent of current proven world reserves and has made an increasingly large contribution to the growth in world oil supplies over the last decade, may well be nearing its peak as an oil exporting area unless some very major and unexpected new discoveries are made. Reserves on the African Continent, at about 100,000m. barrels of recoverable oil, are equivalent to around thirty-five years of output at 1972 rates. Several countries, like Algeria, already seem to have reached their peak. Others, like Libya, are considering levelling off their rates of output, and Nigeria's growth in output is expected to slow considerably over the rest of the decade. According to some recent estimates, another 50 per cent may eventually be added to the reserve figures across Africa as a whole but this would not be enough to prevent the area reaching a plateau of production, probably within the 1970s.

Reserves in other areas are on a much smaller scale altogether. North American resources, at around 50,000m. barrels of recoverable oil, are equivalent to no more than twelve years' production at 1973 rates. Latin American reserves, at about 30,000m. barrels, are equivalent to less than twenty years of output at 1973 rates, and while there are some continuing increases in output from some South American countries, these are being compensated for by a fall in Venezuelan output.

An area of considerable new potential is clearly offered by the North Sea discoveries off the coast of north-west Europe. But, at around 15,000m. barrels of recoverable oil, present proven reserves are still no more than 2–3 per cent of the world's total. Even if the more optimistic estimates of ultimately recoverable reserves of around 50,000m. barrels are taken, this would still leave western Europe with no more than about 7–8 per cent of the world's total and a producing capacity of around 5m. barrels per day, or about 20 per cent of western European oil consumption during the mid-1980s. And the same could also be said of the potential oil output of South-East Asia and Australasia, where Indonesian off-shore discoveries have raised high hopes for growth in production. But reserves in the area, at about 15,000m. barrels, are equivalent to less than thirty years' output at current rates and Indonesian production could well reach its peak at about 3m. barrels per day.

The only countries – on current thinking at least – with the potential to become major rivals to the Middle East as a source of oil exports are the USSR and China. Precise estimates of reserves in either country are hard to come by and figures on China, in particular, tend to be based on surveys before the Communists took power. But, according to recent published estimates, current reserves in the two areas may well be as high as 100,000m. barrels, of which two-thirds may be in Russia. These figures could prove small in relation to the unexplored potential of both China and Russia, whose huge Siberian region looks to be one of the few basins of really major promise left relatively untouched in the world. One recent estimate has suggested that around two-fifths of the world's potential sedimentary acreage is situated in this region, and another estimate has suggested that a further 200,000m. barrels of recoverable oil may lie there undiscovered.

Yet, while Russia has long been an important exporter of oil to the West and could greatly increase its share of the world's oil trade, there are a number of reasons for doubting that it will become a major factor in international oil flows. Like other rapidly industrializing regions of the world, internal demand for oil in Russia and the Communist bloc is rising at extremely rapid rates of 8–10 per cent per annum, and in some individual cases rather more than this. Many of Russia's traditional producing fields on the western side of the Urals are now beginning to go into decline and while the country has made rapid efforts to build up production in western Siberia and other parts, the enormous scale of investment, labour and engineering effort required to produce and transport oil from the most promising regions of remote Siberia is likely to prove a severe drawback to accelerated development for some time.

On most assessments, therefore, the amount of oil available for net exports from the USSR and the Eastern bloc to the West, which rose from around 450,000 barrels per day at the end of the 1950s to around 1m. barrels per day by the end of the 1960s, may well remain stable or even decline during the 1970s. The major part of the increased production in Russia – which has risen from about 3m. barrels per day in 1960 to 8·5m. barrels per day in 1973 and is expected to nearly double to 12–13m. barrels per day by the end of the decade – will be needed for internal consumption in the USSR, which is more than doubling every decade with the growth in energy consumption per head, the growing car population and the increasing pace of industrialization. A similar rate of demand growth is also going on in the Eastern bloc, whose import requirements may reach 1m. barrels per day or more by the end of this decade, so that the amount available for net exports from the Comicon area may well prove small.

This does not imply, as some commentators have suggested, that the Soviet bloc is likely to become a substantial net importer in its own right and hence enter the Middle East for commercial as well as political reasons. Unless something very dramatic goes wrong with its programmes, this is most unlikely. What it has been doing, however, is to seek hard currency exports of oil to the West by asking the Eastern Bloc countries to seek some of their supplies from other regions, particularly North Africa. And this is a policy it could well continue over the future, for political as well as commercial reasons.

On the same score, Russia has also been discussing opportunities recently for Western investment and technological aid in developing oil and gas deposits in Siberia and Asian Russia.

But again it is very difficult to foresee in current circumstances an opening up of Russian oil to foreign help on a scale that might lead to substantial increases in exports over the medium-term future. On oil questions (although perhaps less so where gas is concerned), Russia's aim would appear to be to use Western help in so far as it needs to do in order to keep its output plans on schedule, not in order to create a major export trade in oil.

China's position on oil trade is more difficult to assess. Official remarks suggest that its production rose to some 1m. barrels a day in 1973, largely from the Taching, Shengli and Takang fields. At the same time, China also started to arrange for the first time to export volumes of oil to the West and to discuss possible technical co-operation, in particular with Japanese interests. But, like Russia, China's willingness to develop at a rate primarily directed towards export opportunities must remain doubtful. Exports may increase but overall oil development will probably be kept at a pace which China itself can control and manage.

Theoretical growth in output from the major producers*

| | (m. *barrels per day*) | | | |
---	1973	1976	1980	1985
Saudi Arabia	7·5	10·0	14·0	23·0
Iran	5·9	8·0	9·0	10·0
Iraq	2·0	3·0	5·0	8·0
Kuwait	3·0	3·5	3·5	3·5
Abu Dhabi	1·3	2·5	4·5	5·5
Qatar	0·6	0·5	0·5	0·5
Libya	2·2	2·5	3·0	2·5
Nigeria	2·0	2·5	3·0	3·0
Venezuela	3·4	3·4	2·5	2·5
Indonesia	1·3	2·0	2·5	3·0

* Estimates by Shell based on present assumptions as to reserves.

If this leaves the Middle East holding all the best cards, however, it must remain a very different question whether this area will in fact provide all the oil needed by the world or what conditions it will demand for doing so. The post-war years have seen the producers in an almost constant struggle to bring the international exploitation of their oil more closely into line with their own internal needs and ambitions. This struggle, pursued sometimes with great bitterness and sometimes with surprising good humour, has traditionally been fought against the obstacles of an international oil industry that has controlled the outlets for its oil and a condition of surplus in the market that has encouraged the producers to compete against each other for greater revenue through greater production. As long as there was more oil available than the world needed, the oil producers were forced into an essentially negative fight to ensure that their own tax-reference prices and hence their oil receipts did not decline along with actual market prices. As long as it was a buyer's market, the oil producers needed the oil companies to ensure that there was a steady sale of their products and a steady increase in their production.

The change in trading conditions during the early 1970s from a buyer's to a seller's market has now completely altered these horizons and, in doing so, has enabled the producers to fulfil most of their long-standing ambitions for control of prices and development within the space of a couple of years. In 1970–71, for the first time the producers forced the price of oil up not just in line with market conditions but in answer to their own united demands. In 1973 they followed this up by abrogating the price agreements altogether and taking on the power themselves to set prices in defiance of any market trend.

In the same move, most of the oil exporting countries have taken virtual control of their oil production either through outright nationalization, as in Algeria, Iraq and Libya, or through various types of participation or contract agreements, as in the Gulf. Some of the Arab states participating in the cut-backs also, for the first time, exercised actual control of the destinations for their oil, instructing the companies not to take oil at all to certain embargoed countries, to take additional amounts of oil to other favoured countries and to take other specific proportions of oil output to other less favoured nations. And this has been followed up by a rash of negotiations over direct bilateral sales of oil to certain countries like France in exchange for goods and services, economic development or general political goodwill.

As the producers have taken on more and more economic power, so too have they begun to express this power in political ways as well. After a century or more of colonial or economic exploitation of the West, the Arab producers of the Middle East have forced themselves on to the world's stage, using oil for political ends and, while this may or may not prove to be a once-and-for-all explosion of frustration and new found Arab unity, the whole trend towards direct industrial and financial deals with consuming governments could well bring politics more and more into the day-to-day organization of the oil trade.

The speed and extent to which the producers have changed the face of the oil industry over the past few years, as well as the problems that this has raised, is most obvious in terms of their oil revenues. Oil prices have always been at the centre of producer–oil industry relations and, despite the steady fall in market prices for oil, the producers have been able to maintain a steady rise in earning for a decade and more.

Between 1960 and 1970, the revenue of the major oil exporting countries rose more than threefold from about $2,500m. per annum to nearly $8,000m. per annum, largely through increases in production. The unit revenue meanwhile climbed more gradually from an average 75–80 cents per barrel in the Gulf and about 90 cents in Venezuela at the beginning of the 1960s to around 82–88 cents per barrel in the Gulf and 90 cents to $1.10 per barrel amongst the producers of Africa and Venezuela at the end of the decade.

The potential problems of large accumulations of revenue in countries without the economic infrastructure to absorb it were already obvious from fairly early on in countries such as Libya and Kuwait. Overall, however, the member states of OPEC were still importing more in terms of goods and equipment from the West than they were selling back in terms of oil, while inflation on their imports was rising far faster than the price of their oil exports. And this remained broadly true even after the dramatic expansion of oil revenues between 1971 and the first half of 1973, when oil revenues nearly doubled to around $15,000m. in 1972 and an estimated $22,000m. in 1973, while revenues per barrel rose by more than 50 per cent to $1·40–1·50 in the Gulf and $1.70–2.00 for African and Venezuelan crude oil exports.

The consumer might worry about the potential problems arising from the surpluses in certain countries like Saudi Arabia, but for most of the producing countries like Algeria, Iran, Venezuela and Iraq, oil revenues were still far short of their own needs when the dramatic events of late 1973 allowed them virtually to quadruple oil prices, and their revenues per barrel in two unparalleled decisions in October and December of that year.

These increases have completely altered the picture of producer finances. Total oil revenues on the basis of the prices introduced then could reach well over $100,000m. and much more during 1974. And even if prices should fall in real terms in the face of consumer resistance and the drop in Western oil demand, they seem unlikely to return to anything like their former levels and could, indeed, go on rising if the producers sell their oil directly at higher prices than the tax-paid cost of company-owned oil and if OPEC manages to tie prices to the rates of Western inflation.

At this kind of revenue level, and at the mid-1974 basis of receipts per barrel at $8.50 per barrel in the Gulf and $12–13 per barrel receipts for the 'short-haul' producers, the OPEC members as a whole will gain revenues far beyond their means to spend them and well beyond, in some cases, their ability to manage them in terms of investment outside their countries. Over the short-term at least, the total ability of the producers usefully to spend the money amounts to perhaps $35,000m. per year and, although inflation in the West will increase that figure, inflation could also increase the oil prices at just as rapid a pace. One study by the World Bank made early in 1974 suggested that total OPEC revenues were likely to be of the order of at least $85,000m. in 1974 and, on normal projections of growth in supply and price, that they could rise to around double that figure in 1980 and possibly as much as $200,000m. Individual country revenues reveal even more dramatic figures, with the possibility that Saudi Arabian revenues, reaching some $20,000m. in 1974 – the equivalent of nearly the whole OPEC revenue in the previous year – could rise to nearly $45,000m. by 1980, while Abu Dhabi's revenue, at nearly $5,000m. in 1974, could triple to nearly $15,000m. in 1980 and Iran's revenue, at a possible $15,000m. in 1974, go up to over $30,000m. in 1980.

Quite apart from the strains that this kind of capital accumulation in the producing world might bring to the major money and investment markets of the West, it raises profound and disturbing questions about whether the producers will be prepared to go on raising oil production and thus revenue on this kind of scale and where they will decide it best to invest their money. In one blow the producers have brought forward to 1974 all the worst fears about what might happen by the end of the decade in terms of financial and political power and the problems of unrestricted growth.

It will be the next few years that will see the answers develop to the central questions posed by the rapid changes of the last few years and the chances are that they will emerge in a much more varied pattern from country to country than has appeared before. Both the central push on prices and participation through OPEC together with the heady air of Arab unity that the oil cut-backs have given the Middle East and African producers have tended to suggest an appearance of co-ordination and cartel operation that has never been truly there. The strength of OPEC has come not so much because the producers have agreed amongst themselves to act as a united body but simply because, in the changed circumstances of tight supplies and a seller's market, any one producer has the power to disrupt supplies and achieve particular demands. In an atmosphere of constant competition between the major producers not to be outdone in any of their struggles with the oil companies, other countries have then tended to follow on with similar demands. It is this kind of momentum that the oil companies have fought most strongly to control – with only partial success – over the last few years, and in the past it has been one of the most important dynamic influences in producer relations with the industry.

Now that the producers have gained many of their common aims – unilateral control of prices accompanied by dramatic increases in revenues, as well as more direct control of production and a growing interest in using oil as a central part of diplomatic and commercial relations with the West – the basic differences in their individual outlooks and situations is becoming more apparent. Most obviously these concern politics. There is a wide gap between the radical producers of the Mediterranean such as Libya, Algeria and Iraq, and the more conservative, pro-Western royalist régimes of the Gulf, such as Saudi Arabia, the United Arab Emirates and Iran. On another level, the cutbacks and embargoes of 1973–74 have also sharpened the political differences between the Arab and non-Arab members of OPEC as well as producing some surprising conflicts in approach such as that between Iraq, with its desire to attack Western assets in its country, and Saudi Arabia, who preferred the route of destination controls as the political weapon.

But these political differences also reflect more fundamental distinctions in economic terms. On the one hand, there are those countries with very considerable potential for growth but without the infrastructure or the population to absorb massive increases in revenue. Most important of these countries is Saudi Arabia, with its vast reserves of around 150,000m. barrels of recoverable oil – some three times larger than those of any other country in the world and sufficient to support a production of some 20m. barrels per day or more without straining its resource limits – but with a population of only some 7–8m. people, many of whom continue a nomadic life, and with very limited opportunities for spending the vast sums it is now getting from oil without indulging in a conspicuous consumption that could have disastrous results on its social stability.

No other country is on quite the same level as Saudi Arabia. But another state with similar problems of excess revenue but still potential for growth is Abu Dhabi, with a population of less than 200,000 but an eventual production potential of possibly 5m. barrels per day. There are also countries like Kuwait, with a population of less than 1m. and a revenue expected to reach $4,000–5,000m. in 1974, and Libya, with a population of 2m. and an estimated revenue of $7,000–8,000m. in 1974, neither of whom seem to have great potential for growth but both of whom have little need for even their current revenues.

At the other end of the scale there are countries with almost

unlimited needs for oil revenues to fund their economic programmes and feed their large populations. Some, like Iran and,
even more, Iraq, have at least some potential for further growth.
Iran is already planning to increase output to 8–9m. barrels
per day by 1976 compared with 6m. barrels per day in 1973,
while Iraq, whose territory still remains relatively unexplored,
could go to 5m. barrels per day by the same time and perhaps
even more by 1980. Both have large populations – over 30m. in
Iran and over 10m. in Iraq – and both have every incentive to
go on increasing production and raising prices to fulfil their needs.
Of less certain potential is Indonesia, but with a population of
128m. people it too has urgent need to develop as rapidly as
possible. And in the same general category must also come
countries like Algeria and Venezuela, with large populations
and ambitious development programmes based on oil revenues,
but unfortunately limited in their production potential.

All these countries inevitably must have different views on how
they wish oil and oil prices to be developed over the future in their
own best interests. Some, like Iran, Iraq, Venezuela and Algeria,
have considerable prospects for developing balanced economies
outside of oil and, in Iran's case at least, the major problem must
be to use oil to finance this potential even at the expense of running
down reserves and to ensure the maximum price possible for oil.
Other countries do not have the same prospects for internal
development. If they have limits to reserves like Kuwait, they
must naturally worry about sustaining their natural resources for
as long as possible while, if they do not, like Saudi Arabia, they
must worry about the impact that high prices must have in
driving consumers away from oil use, leaving them still with oil
but without markets over the long-term future.

This is not to suggest that OPEC as an institution is necessarily
doomed to disintegrate over the future. It has proved a useful
and an effective co-ordination mechanism in the past for expressing the producer views not only on prices but a wide range of
other aspects of relations with the oil industry as well. So far there
are no signs that the producers as a whole wish to promote conflict
with the consumers either by using their financial reserves to
upset Western financial institutions or to create an atmosphere

of continual instability on prices. Indeed, in the light of reduced demand in the West and the possible development of alternative fuels to oil, the producers may well decide that it is in their interest to create new price arrangements directly with the consumers which will guarantee them against inflation and enable them to plan their own long-term budgets in an atmosphere of reasonable long-term stability. At the same time they may also wish to come to some arrangement for the investment of their surplus funds which guarantees them the same insurance against inflation and short-term upsets in currencies.

OPEC, with its established staff and co-ordinating role, is the obvious means for negotiating such arrangements as well as continuing to be an essential office for collecting data, and providing advice and a useful forum for the exchange of views between oil and finance ministers of the producing worlds. Whether it will prove to develop beyond this depends very largely on whether the producers themselves continue to have enough common interests in such questions as direct producing area-to-consuming area negotiations of trade terms with the European Economic Community or wider consumer organizations.

Yet, even on the price question, there are indications of emerging conflicts between the producers, with countries like Iran anxious to maintain prices at as high a level as possible and other states like Saudi Arabia less interested in the maximum possible revenue per unit and distinctly nervous about forcing the pace too harshly with the consumer. Much of OPEC strength as an organization has come from the common interests of its Gulf members in the past, but this is being dissipated both by the entry of new members from Latin America and elsewhere, with entirely different backgrounds and approaches to the Gulf states, as well as by the emerging difference on economic as well as political questions between Iran and Saudi Arabia. The events of 1973–4 have already seen a switch in the leadership of the producing world away from the radicals, such as Libya and Iraq, who have made so much of the running during the last five years, to Saudi Arabia, who has by far the greatest weight in oil production terms and is more and more expressing it on the international stage.

Saudi Arabia alone now has the power to force prices up or ensure that they fall simply by either restraining the rate of growth in its output or letting it expand as rapidly as possible, and one of the central themes of the remainder of this decade is almost bound to be the way in which Saudi Arabia exercises this power and how its policies develop in relation to those of Iran, with its greater economic and political power in the Gulf, and possibly with Iraq as well.

The differences in economic circumstances between the major producers, and the relative weight of Saudi Arabia, Iran and other states, is also likely to have an important bearing on the way in which producer country attitudes to other oil issues develop. There are several courses which almost all the producers are likely to pursue. One is the move towards greater and greater direct control of the production and disposal of their oil. Whether participation survives as a means of ensuring stable changes or not – and it remains dubious that it will at this stage – all the producers will now be anxious to ensure that they have agreements giving them a permanent settlement that will not quickly look outdated as the participation agreements have done. But again this may involve different arrangements in different countries, some of whom, like Algeria, Iraq and Iran, have the experience and organizations to take on full management of production and sales in a way that others, like Libya and Abu Dhabi, have not. Thus, while some states may prefer complete nationalization, others may be satisfied with a long-term contractual relation with the oil companies on the lines that Iran has negotiated with the Iranian Oil Consortium.

Nor does this necessarily mean an end to the presence of the traditional concessionaire companies in the producing areas. True, many of the traditional reasons for maintaining their presence – the guarantee of markets, their vulnerability to pressure on prices and their investment – have now been weakened, if not been made obsolete, by the changes in market conditions and rapid increases in revenues. But they still retain the attraction of experience and management which even the most radical countries like Algeria, Venezuela and Iraq may prefer to use, even if it is on a more purely contractual relationship. In an age, too, when no one can predict for certain how long the market will continue in favour of the seller rather than buyer, some producers, particularly the less-experienced in the Gulf, may prefer the continued presence of the majors as a guarantee that a certain proportion of their trade will be under stable price and marketing conditions.

Another area where all the producers are likely to develop as rapidly as possible is in substantial investment in oil-related projects within their own boundaries, through the construction of oil refineries, petrochemical and fertilizer plants. Attempts along these lines in Kuwait, Iran and elsewhere have not been notoriously successful in the past. But the change of market conditions, the need to develop industries at home and the prospect of large oil incomes over the future are all bound to give the development a major new boost. Most of the producing states, especially in the Gulf, are now planning large-scale export refineries, often in partnership with the oil companies, and most of the producing states are also evolving ambitious plans to create the most logical next steps towards the construction of ethylene, ammonia, polyethylene, aromatics, PVC and methanol, while several of them also have plans for the manufacture of more sophisticated chemical products such as pharmaceutical and synthetic rubber as well as manufactured goods.

Theoretically, this kind of investment could radically alter the nature of the world's oil trade, which has tended to move increasingly towards the refining of oil at the market-place and is now threatening to swing back to the refining of oil in the place of production. But producer-government participation at present is small, at perhaps some 1 per cent of the world's refining capacity and much less of the world's petrochemical production. Even at the most ambitious, it seems unlikely that the producers could raise that share to more than 5–7 per cent within this decade, and their ability to do even this will tend to be constrained by limitations on the part of contractors and suppliers of equipment to deal with such expansion. A weakening of prices for chemicals and the tendency of even developing countries to insist on refining their own products from oil in order to reduce the balance of payments cost and increase their own development potential could also act as restraining factors on this development although it will not prevent it happening on an increasing scale.

The same kind of limitations and uncertainties also arise in any consideration of the producing countries' likely policy towards investment outside their own countries and in their relations with the consumer. Participation 'downstream' in the transport, refining and marketing of oil has long been something which various oil experts within the producing world have spoken about. Some countries like Iran and Algeria, with experienced oil companies of their own and with ambitions to become a growing part of the industrialized world, are already developing along this direction and will probably go further. But there are drawbacks in terms of the 'hostage' that it provides with the consumer, the lack of flexibility on pricing and supply that it induces and the possible objections of the consuming government. These factors could well induce other countries to be less ambitious in this direction.

Bilateral trade relations involving oil provide similar quandaries for the producer. They undoubtedly have the attraction of guaranteeing reverse investment by the consuming countries in the kind of manufacturing development that the producers want and they could also, at a time of uncertainty in market conditions, provide stability of markets. But, by their nature, they equally reduce the flexibility of the producing countries and raise the inevitable problem of what is good for one side is all too often bad for the other, whether it be on prices or trade in other goods. For many countries it may well provide a useful form of commercial and political relationship covering a proportion of their oil trade. Some, like Libya, may even see this as covering the major part of their oil exports. Others, however, may prefer greater flexibility in their trade patterns and many may well withdraw from tying themselves too closely to any one relationship lest it go sour on them, as the Franco–Algerian agreements did.

Finally, there are the strong attractions to the producers of investing and developing close relations based on oil or the revenues from oil with the developing world. The Arab oil states form but a tiny proportion of an Arab world all too often marked by poverty and a backward state of economic development. Several funds have now been set up by the producers, both on their own account and as a united group, to invest in these areas, while the Libyan and Saudi Arabian jostling for position in Egypt has illustrated the attractions for the richer Arab oil states of using their oil revenues to extend their influence in other parts of the Muslim world. At the same time, interest has also been shown by countries such as Iran in developing close economic ties with other parts of the developing world in both Africa and Asia.

A certain consciousness of their responsibility to the developing world, coupled with a rather more self-interested sense of the investment potential there, did mark some of the actions of the producers during the 1973–74 crisis. Further development in this direction provides a strong possible direction both for producers like Iran, anxious to develop as major economic powers in their own right, and others like Libya or Saudi Arabia, with a need to express their financial wealth in tangible form of foreign influence. But the producers as a whole have yet to show a willingness to gain this influence at the expense of lower financial return to themselves, and whether a form of 'neo-colonialism' by the producers will prove acceptable to the developing world in Africa or Asia remains to be seen.

Overall, therefore, the future presents as many uncertainties and as much indecision to the producers as it does to the consumers and the oil industry. Most of them have risen to wealth and power at an extraordinarily rapid rate. Quite aside from the internal tensions this has brought, it has also left them with very real problems of how to arrange their development to ensure the best interests of their own countries in a world that may move rapidly away from imported oil as a major source of energy.

For some producers such as Iran and Iraq the question will probably induce as rapid a rate of oil development as possible in order to ensure speedy industrialization in other directions within their boundaries. For other countries, like Saudi Arabia, the answer may not be so easy as they consider whether to restrict production to the level at which they can usefully use the revenues and hence prolong oil production as far forward as possible or to go all out to produce as much oil as they can sell and use the money in long-term investments abroad.

Libya and Kuwait, both of whom have limited reserves, have already answered the question by adopting a policy of restricting output. Saudi Arabia and Abu Dhabi, both of whom could achieve substantial increases in production but have little financial incentive to do so, have yet to make up their minds. There are government officials and politicians in both countries who are urging such a restrictive course on economic grounds just as there remain political as well as economic reasons for

The major oil producers

Country	Population (million)	1973 production (m. barrels estimated)	Oil reserves	Ratio of reserves to 1973 product.	Estimated 1974 revenues* ($m.)
Saudi Arabia	7·74	2,800	140,750	1:50	19,400
Kuwait	0·91	1,148	72,750	1:63	7,945
Abu Dhabi	0·20	551	24,000	1:44	4,800
Qatar	0·08	203	6,500	1:32	1,425
Libya	2·08	773	25,500	1:33	7,990
Iraq	10·07	689	31,500	1:46	5,900
Iran	30·55	2,190	60,000	1:27	14,930
Algeria	15·27	378	7,640	1:20	3,700
Nigeria	58·02	730	20,000	1:27	6,960
Indonesia	121·63	475	10,500	1:22	2,150
Venezuela	10·97	1,230	14,000	1:11	10,550

* Estimates made by the World Bank in January 1974.

continuing to raise output. And the problems of the producers are made all the more severe by the rate of inflation, the fall in Western demand and the currency upheavals which threaten to reduce substantially the real value of oil in the market and the real value of their monetary revenues from it.

The question they face is of crucial importance not only to the producers but also the consumers. Without substantial growth in supplies from the Middle East, the consuming nations will face considerable difficulties over the shorter and medium-term before other fuels can come into play. With increased supplies, especially from Saudi Arabia, their problems could be greatly reduced. Yet in an age when the Israeli question has removed many of the last traces of direct Western power in the Middle East, there are few ways of forcing the producers to raise output to meet the West's needs. Economically it can be argued that oil is now at the height of its commercial influence and that to leave it in the ground at this stage is to court disaster when the consumer finally moves away from relying on it. Politically, too, it can be argued that the interests of the conservative régimes of the Middle East are too closely linked with the West for them to starve it of its essential oil. But it could equally be asserted that the low production cost of Middle East oil will always make it a force in the market, that increased output will only create social tensions in countries unable to absorb the revenues and that more gradual development is in the greater interests of future generations of Arabs.

In an age, after all, of economic nationalism, when Canada, Australia, Norway and even Britain are all talking in terms of restraining their oil development within the confines of their own national needs, it is becoming increasingly difficult to argue that the major producers should act differently. After a decade of constant struggle to gain the right to control the exploitation of their own natural resources, smarting under the real or imagined rule of Western 'neo-colonialism', the producers will take some persuading to use that control primarily in the interests of others, unless very sound reasons of self-interest can be presented to them.

CHAPTER 23

Consumer intervention

'In fact, the oil companies are not only freelance operators, they are actually purchasing agents of the consumer countries and it is not unreasonable to ask whether the interests of these companies ever have been or at least are any longer identical with those for whom they get the oil.'

*Dr Paul H. Frankel**

The history of the oil industry has been written for so long in terms of the relations between the major oil companies and the producers that the consumer himself has often tended to get relegated to the background, at best a shadowy figure and at worst an almost irrelevance. Yet one of the most important of all the developments to come from the climactic events of the early 1970s could well be the emergence of the consuming countries as principal actors in the drama as it unfolds over the next decade. For all the problems of divergent views and hesitation on the part of the consumer, the Washington meeting of the major industrialized countries in February 1974 marked the first tentative step towards a real co-ordination of views among the consumers and the creation of the machinery to produce common policies. For all the obvious flaws of rivalry between consumers, the direct negotiations now being pursued between producer and consumer on trade, finance and industrial investment could prove to be the beginning of a direct relationship between the two sides that is both inevitable and quite possibly long-lasting.

Previous attempts by the consumers to intervene in energy and oil questions do not give rise, it is true, to any great degree of optimism about their ability to manage things in the future. Of

* In a speech entitled 'The Current State of World Oil'.

all the major consuming states, perhaps only France has ever produced an entirely logical and considered approach to the oil trade, based on a *dirigiste* control of the market, in which an orchestration of company shares of sales has enabled the government to manage refinery investment and crude oil sources. This has been accompanied by a foreign policy conscientiously devoted to establishing a favoured position in the Arab world, deliberately setting itself up as an alternative to United States' influence and seduously promoting bilateral oil arrangements with Algeria and, more recently, Iraq, Libya, Iran and Saudi Arabia. At the same time the French government has also supported and subsidized heavily French companies in an effort not only to create a strong, technically-orientated oil industry of its own, with the full paraphernalia of educational and research establishments, but also other energy technologies from nuclear to solar and tidal power as well.

But even French policy has been marred too often by an obsessive desire – reflected also in other areas of its policy – to 'do down' the United States; by an almost proprietorial attitude towards Algerian oil, which produced its reaction in the Franco–Algerian conflict of 1970–72; and by a tendency, common to all consumers, to try to have it both ways at once by playing off foreign oil companies against French oil companies and by playing the semi-state-owned CFP group against the wholly state-owned group, Elf–ERAP (now merged with the state-owned Aquitaine oil and gas company). Nor, in the final analysis, has French policy yet proved more obviously successful in gaining security of supply or cheap energy than the policy of other more *laissez-faire* countries.

Other consumer nations have followed a less consistent and considerably less ambitious policy towards oil in the past. In the immediate post-war years, when the most urgent need was for economic reconstitution of the broken economies of Europe and Japan, the main emphasis of oil policy was turned to reducing its balance of payments impact on the importing countries by increasing refining capacity in the market, reducing imports of products and to market-related transfer price for oil in place of the artificial United States-related posted price which to prevent the concentration of profit in the headquarters of the integrated companies.

If there were worries about the impact of oil on security of energy supply and its effect on the traditional coal industries of Europe and Japan, actual policies to control oil's development in the market-place and to ensure the flexibility of alternative sources of energy tended to be sporadic, ineffective and often short-lived. Taxes on fuel oil were introduced in countries like the United Kingdom on the grounds of aiding more expensive coal to compete. Public services such as the electricity generation authorities were 'influenced' in their choice of primary fuels. But taxes on oil became too easily a method of raising money rather than influencing competitive prices and, even though continually increased, could not make up for the more rapidly escalating cost of labour-intensive coal or high-technology nuclear power. Defence of coal tended to become a defence of employment in the coal industry, bought, as in the United Kingdom, at the expense of any really sustained investment in the industry, and

Growth in the West's dependence on oil, 1960–70

Area	1960		1970	
	Volume (m. b/d)	% of total energy	Volume (m. b/d)	% of total energy
OECD Europe	0·4	32·6	12·4	59·6
North America	10·2	44·9	15·3	43·3
Japan	0·6	36·4	3·8	71·7
Rest of World	5·5	22·4	10·7	31·8
Total World	20·3	33·2	42·2	43·9

Source: OECD Oil Committee

the occasional doubts about oil's vulnerability to cut-offs that occurred during the closures of the Suez Canal proved short-lived as alternative oil sources were found and the practical problems of developing 'crash' programmes of nuclear expansion were encountered.

Under these conditions consumer government policy towards oil tended to move away from attempts to control its growth in the market to efforts aimed at gaining greater national participation in the oil flows. Countries like the Netherlands and Britain, of course, had major oil companies of their own, and for them questions of whether the international oil companies effectively defended their national interests were correspondingly less pressing. But other countries had obvious doubts.

In Europe, France was the major proponent of pushing for a national alternative to the major oil companies. But other countries at least made efforts to ensure that a greater share of their home markets was held by national companies and that national oil companies were encouraged to go out and search for oil themselves through new exploration agreements in the producing world. Italy's ENI proved particularly active in both directions, although how far Mattei's ambitions genuinely reflected national policy as opposed to his own ideals is less clear. Spain developed both a marketing monopoly through CAMPSA and an amalgamation of national interests for exploring abroad in Hispanoil, and even Germany, the bastion of *laissez-faire* economics, attempted in the 1960s to see that a proportion of its own market remained in national hands through controls on sales of equity in local oil companies, while at the same time encouraging the formation of a consortium of companies to look for oil abroad in the Deminex group.

Similar ambitions, although expressed in rather more subtle forms of financial state management, also prevailed in Japan, whose government right from the beginning of the post-war period insisted that only 50 per cent of shares in refining ventures could be held by foreign companies, buttressing this by a series of foreign exchange regulations. As oil became more and more a predominant fuel in the market, Japan, like the European states, looked to more effective control of the market through the 1962 petroleum legislation. This strengthened government control of prices and refinery investment within the country, enabled it to do far more to encourage the Japanese search for crude oil sources abroad through fiscal incentives and also gave a stronger hand to the government in developing bilateral trade deals with the producing states – all lines which were subsequently followed up with the creation of a Petroleum Development Public Corporation

in 1967, the development of trade ties with a large number of
Middle East countries and the increasing investment by Japanese
consortia in overseas exploration and production ventures
climaxing in the purchase of a 22·5 per cent share of the Abu
Dhabi Marine Areas concession from British Petroleum in 1972.

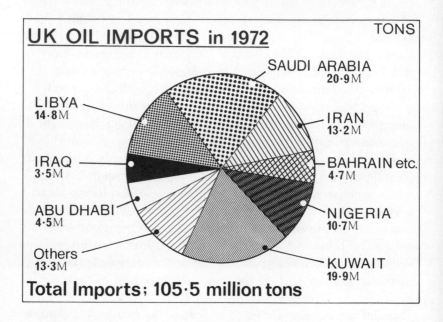

FIG. 16

The success of these various attempts in Europe and Japan to
ensure greater national control of oil sources and sales, although
far from entirely fruitless, was never as great as their proponents
had hoped. In terms of local market, most countries at least
managed to confine the growth of the international oil companies
and to ensure that a reasonable proportion of sales went through
national companies. By the end of the 1960s, 45 per cent of the
Japanese market was in the hands of local companies compared
to around 30 per cent in the previous decade, while in Italy,
Germany and elsewhere the share of sales held by local concerns
rose to at least a quarter.

But crude oil sources tended to remain firmly in the hands of the major oil companies. The aim of many consuming countries in setting up oil companies seemed too much like creating further international majors in a different guise to provide great appeal to the producing states or to have much chance of success at a time when the great reserves of cheap oil in the Middle East and Africa were largely in the hands of the international majors and United States' independents. The new national companies had to struggle with less prospective and more costly acreage in the world, and too often ended up with more expensive oil at less advantageous concession terms than the majors held. Some three-quarters of the crude oil imported into Japan continued to come from the major oil companies, while a similar proportion was also true in most countries of Europe, although with greater reliance on United States' independents in those countries taking Libyan oil. As long as oil was in plentiful supply, the effort of the national oil companies was bound to be a marginal one.

Nor were attempts in Europe to create an international force to achieve greater security of oil supply through the European Economic Community of any greater overall success. Although the Commission in Brussels produced a steady stream of reports and recommendations designed to promote a more orderly and more closely supervised oil market, to develop oil-orientated trade agreements with the producers, and to protect indigenous energy sources and invest in alternatives, too often its proposals ended up as they had started; interesting advice rather than genuine policy moves.

Too many of its recommendations were the product of a bureaucratic desire for uniformity and centralization to have much real appeal to a group of nation states each of which had very different views on energy matters. The non-coal producers were hardly anxious to help subsidize the major coal countries. Research and development of nuclear technology and uranium enrichment became the subject not of a common effort but competing national programmes and individual ambitions for local contractors and suppliers. On oil questions there were deep divisions between the free market ideals of Germany and the *dirigiste* policies of France; between those countries like the Netherlands, with major oil companies of their own, and those like Italy without; between those like France who wanted to give favoured treatment to particular crude oil sources, notably from Algeria, and other countries who wanted to diversify sources as much as possible and, finally, between those countries who wanted to move in co-operation with the United States and the United States-based oil companies and those, like France and to some extent Italy, whose policies were clearly designed to undermine the dominance of American economic foreign policy and the position of its oil groups. There was development of individual questions of storage, harmonization of duties, etc., but the idea of a common European energy or oil policy remained virtually stillborn throughout the 1960s.

Consuming governments should not be blamed too much for these failures. Politicians are never very good at undertaking long-term decisions unless there is a very obvious, and publicly-accepted, short-term need to do so. And this there never was during the post-war decades. Most countries emerged from the Second World War only too grateful to gain the benefits that oil could bring to their industrial revival to worry too much about the long-term consequences or the routes by which it came.

Even had the consuming countries wished to curtail the growth in oil, there were limits to how far they could go without imperilling industrial costs at a time of rising international competition for the sale of the manufactured goods, without drastically intervening in the consumer's right to mobility with the increasing use of the car and without a dramatic intervention in the market choice of fuels. As long as oil was cheap and in abundant supply, its rise as the dominant market fuel was probably inevitable and certainly economically attractive to the final end user.

Equally, the necessity to take drastic action to ensure national control of oil flows was far from proven in the 1960s. Germany, with its free market policies, undoubtedly gained cheaper fuel costs over the 1960s compared to those countries with more interventionist policies like France. And the benefits gained by France in terms of security of supply were rather weakened by the experience of the Franco–Algerian conflict. Through two successive crises caused by the closures of the Suez Canal, the oil industry managed an effective redistribution of available resources which seemed a great deal more successful than if the countries involved had been tied to particular sources of crude oil on their own account. For all their faults, the oil companies did seem to provide an effective buffer between consumer and producer and one which relieved the consuming governments of any need to raise funds on their own or become deeply involved in price and supply negotiation.

Against this background of consumer passivity the oil crises of the last few years have proved a chastening experience. For the first time, the industrialized West, and the developing world, has been made to face the reality of its growing and almost unwitting dependence on imported oil. Despite the much-vaunted ambition of most governments in the 1960s to achieve the full flexibility of a competitive four-fuel economy of oil, gas, coal and nuclear power, nearly all have found themselves with oil not only forming the major part of their energy consumption but also taking virtually a monopoly of many sectors of the industrial and domestic fuel market.

In Japan's case, imported oil, which supplied 36 per cent of its primary energy consumption in 1960, now accounts for almost 75 per cent. Western Europe has moved from a position where oil accounted for less than a third of its energy needs in 1960 to nearly two-thirds in 1973. Almost everywhere coal has been allowed to go into nearly irreversible decline while nuclear power has slipped, partly for problems beyond the scope of government control, well behind schedule. Gas, although the fastest growing fuel outside America, has been limited by the amounts of supplies available and the rigidities of its distribution system. If nothing else, the last few years have forced governments seriously to reconsider their own role in the energy market-place and their policies towards oil.

At the time of the Tehran and Tripoli price agreements, most consuming governments had probably accepted the fact that the oil companies no longer formed an effective buffer between their interests and those of the producers, and that the companies were equally vulnerable to concerted or even individual action by the producers. Even then, few countries were ready to intervene directly into relations with the producers. There were obvious fears over the insecurity of supply that this might incur, as it had done with the Franco-Algerian agreement and was to do again in the Libyan take-over of British Petroleum. Few countries had any real experience of the oil industry as an international business and there were understandable fears of the unknown. Now the consumer really has no choice. Not only have the producers stripped the final vestiges of control over prices or supply away from the oil companies by their actions in 1973, but the issues that they have raised by involving oil with politics and by introducing such massive price increases are issues which only the consuming states themselves can hope to solve.

Having been reluctantly dragged to the conference table to represent their own interests, however, the consumers are still far from deciding what they will say or just how they will say it. No more than for the producers does the mere fact of taking over some of the role of the international oil companies necessarily provide the consumer with the answer to what to do next. Differences between countries on political and economic grounds remain as strong as ever and may, indeed, even have been accentuated by the recent problems of supply. Some consuming nations, like Japan, have less opportunity for a rapid move away from oil than others, like the United Kingdom with its North Sea reserves of oil and gas and its comparatively large coal industry. Some countries, like Germany with its large reserves of foreign currencies, are better able to deal with the price hikes of the last few years than others, such as the United Kingdom and Italy, which were already in currency problems before oil became a major factor in straining their balance of payments. Some countries have political and military influence in the Middle East, others have not, just as some have a great deal more to offer the producing states in the way of technical help than others.

It is with these continued differences in interests that the consumers are now approaching the most fundamental issue raised by their entry into the oil trade: whether they should throw in their lot with each other and pursue policies and negotiations as an internationally co-ordinated body of major consuming areas; or whether they should seek to manage as best they can on a national basis, seeking bilateral arrangements with individual producers; or whether they can, as they have done in the past, play it by ear and attempt to have a measure of both individual and international co-operative action.

The most immediate problem facing them, and the one that could well induce the greatest degree of international co-operation, is the monetary question. The sheer size of the price rises imposed by OPEC in October and December 1973, coupled with the producers' threat to let prices move with the market, took most consumers completely by surprise and there are still strong differences of opinion about how best to deal with it.

In a sense the problem is not a totally overwhelming one. Even at the peak levels reached in mid-1974, a level of $9–12 landed cost for crude oil is something that most industrialized markets can bear in terms of its direct effect on industrial costs. A high proportion of the final price of oil remains that of consumer tax, which could be reduced if necessary, and the sudden jump in prices – great though it is – is still not enough to do more than reverse the previous trend for a fall in the factor cost of energy as a proportion of most countries' GNP.

At the same time there are certainly some reasons for believing that the producers, in pushing prices quite so high, have brought them to their long-term peak. Some commentators have suggested that, at mid-1974 levels, they were beyond their peak and likely to fall fairly quickly in answer to price-induced rises in supply and reductions in demand. Even if this does not turn out to be the case, however, they are now well within the range of the costs of alternative fuels such as tar sands, shale and even synthetic oil produced from coal, and as such are unlikely to rise much more in real terms over the future, and may indeed fall when compared to general inflation.

But the price increases have nevertheless caused a serious problem of adjustment in balance of payments terms for the consumers. Although rising producer-government expenditure on Western goods and services may ease this problem with time, its immediate impact is to throw virtually every oil importing country, and especially the developing world, into serious imbalance on its terms of trade as well as threatening to cause considerable strains on Western financial institutions by so dramatically switching funds from the consumer to the producer. The suggestions of the ways in which this problem of producer surpluses and consumer deficits could be sorted out are legion, ranging from the idea that the IMF should borrow from the producers and recirculate the money to finance the deficits of the consuming nations, to proposals to create new consumer–producer funds for investment in the developed and developing world, to arguments that the money could be managed and recirculated simply in the eurocurrency markets or by the producer-government purchase of gold and/or other commodities.

Whatever the particular techniques for ensuring that this transfer of funds does the least possible damage to the world's international monetary system, any solution will require a degree of co-operation between all the major consumers, as well as between the producers and consumers, if a disastrous descent into competitive devaluation and a succession of trade wars is not to result as each country looks after her own. And the need for action at an international level is only made more urgent by the appalling predicament in which the oil price increases have placed the developing countries of Africa and Asia – most of whom are highly dependent on oil imports to fuel their industrial growth but are all too vulnerable to the balance of payments strain both of higher prices and the inflation expected in the cost of the Western goods that they import.

Yet even on the financial issues, there remains a strong temptation for those countries with strong reserves of gold or foreign currency and with relatively secure supplies of crude oil to attempt to raise loans on their own to see them through the crisis. And this conflict between an international and national approach is likely to become even more apparent when the consumers consider the question of how best to ensure future supplies from the Middle East and how best to meet any future supply crises.

Here again, the arguments for international co-operation are strong, and the experience so far is not a particularly encouraging one. In their first real test of consumer co-operation on oil during the supply cut-backs and embargoes of 1973–74, Europe and the major consuming nations as a whole presented an extraordinary picture of panic, divisiveness and indecision. Pleas by the oil companies for a government-backed allocation scheme failed before consumer fears that this would only antagonize the producers further, while attempts to share supplies on the part of the companies themselves met with bitter resistance from those who felt themselves better protected.

This may have been understandable enough. The consumers were caught off guard and, in the event, they were probably right in thinking that formal allocation would only have made the situation worse. But the experience did show that, without government co-operation, the results were all too likely to be rapid rises in prices and serious supply disruption as each individual country scrambled for available supplies. For some at least – most notably the United States – the obvious conclusion was that the consumers would have to organize themselves to provide both a countervailing force to the producers and to ensure a reasonable degree of stability in the market by refraining from competing between themselves. Such a consumer body representing the major importing nations of Europe, the United States and Japan could, it was argued, tackle not only the financial problems raised by the oil crises but also ensure the necessary co-operation and investment in the development of alternative fuels and enable a secure agreement on future supply and financial questions to be worked out with the producers.

Logically attractive though this approach may be, however, there remain considerable obstacles to its achievement, most notably the divisions between the consumers. More vulnerable to producer cut-offs, Europe and Japan continue to fear that any co-ordination of consumer interests will simply look like confrontation to the producers, and hence encourage action against them in a battle for power in which the consumers must in the short-term hold few cards short of military action. And if European governments have found it difficult to support each other at a time of crisis, there are even more forceful arguments against their getting together with the United States and Japan, while the need for such unity is obviously felt more by those in the most vulnerable position on oil supplies than those who feel – as some do – that they can arrange better deals on their own.

In the absence of real international agreement on supplies, consuming countries are almost bound to seek other national arrangements to make their weight more directly felt in their oil flows. And from this point of view bilateral trade arrangements with the producers on a government-to-government basis offer a number of attractions. For all their obvious faults in such arrangements do give some security of supply in that the producer is tied to the consumer by technical and trade deals. Technological and industrial assistance from the West is something which some producing countries at least want to develop and hence must attract the attention of the consumer if he is anxious about supplies. By setting off, however indirectly, oil flows in one direction against flows of trade and assistance in the other, the deals can do something to relieve the balance of payments burden imposed on the consumer by high oil prices. And they do provide the producer with an incentive to keep the oil flowing.

Pre-crisis estimates of oil import requirements in the West

	1950	*1960*	*1970*	*1980**	*1985**
	(*in million barrels per day*)				
North America	1·0	2·2	3·3	14·0	18·0
West Europe	1·4	3·6	12·8	19·0	24·0
Japan	0·2	0·6	4·2	9·0	12·0

* Predictions.

Just how far such state-to-state arrangements will become a primary influence on oil trade is something, however, that remains extremely uncertain. For the moment, most consumers feel highly dubious about the advantages of directly tying oil supplies in bilateral deals. Many recognize the inherent dangers of relying too closely to limited sources of supplies which could become the object of further political action by the producers. There is also a growing understanding among consumers – reflected in the limited accord reached by the major oil importing regions at Washington in early 1974 – that competition between themselves for available resources, particularly if expressed in competitive bilateral deals, could have ruinous effects on prices and supply stability. And there is also a recognition by most governments, again reflected at Washington, that co-operation with the United States, however distasteful to some, is a necessity in the light of America's potential dominance of oil supplies through its own imports and its ability to outbit its competitors for available supplies.

On the other hand, there is still the reluctance of governments to produce anything like a cartel of consumers; the suspicion of Unites States' motives and the inability to get over fundamental differences between even allied states in Europe. Government to government negotiation on trade question, which was once an isolated effort by particular countries, is now a route which almost every consuming nation is exploring through diplomatic visits to the Middle East. France has already gone far down the road. Italy could follow in her wake, while Japan, and to some extent Germany and Britain, are likely to pursue such deals when and where they see the industrial and technical opportunities suit them.

But there is likely to be severe limitations to the extent to which countries will be prepared to tie up directly more than a minority share of their oil supplies in this way and the extent to which they are prepared to do it in competition between each other. Much will undoubtedly depend on market and political conditions. If prices continue to ease and supplies seem less tight – the situation prevailing in the spring of 1974 – then much of the pressure will come off the movement to bilateralism. At the same time, if Washington or other conferences result in broad agreements between the consumers as a whole and the producers, or between the EEC as a whole and the Gulf, then the consumers will probably accept a degree of international co-ordination of their individual deals. But if attempts at international agreements fail before the divisions between the consumers themselves or the hostility of the producers, and if supplies once again become dangerously tight, then the tendency will be for every country to act on its own, whether it means pre-empting other people's supplies or not. And if the producers unsheath the oil sword for political purposes again, then it is all too likely that the consumers will once more end up in an indecisive attempt to do both things at once.

Much the same considerations will also apply to the consuming governments' policies towards intervention in their domestic oil markets* and towards control of the oil companies. Undoubtedly, countries will pursue the attempt to raise the share held by national companies in their own markets, both by squeezing foreign oil companies with financial regulations and by direct control of mergers, refinery investment and distribution systems. Unquestionably, too, governments will want a much greater direct supervision of the international oil companies' handling of crude oil flows, the prices at which it is sold and the profits which are made out of it. Too many governments felt themselves hopelessly dependent on oil-company information during the crisis of 1973–74 for them not to try to improve their control in the future. But whether this control takes place at purely national or at international level will again depend on how far international co-operation on general oil questions extends. If some general co-operation between consuming areas is successfully promoted, as it could well be on matters of general supervision and co-ordination of supply schemes, then the question of whether the international or national oil companies act as the agents or continue to manage oil flows becomes less relevant. But if the situation does degenerate into an 'each man for himself' atmosphere, then the tendency will be for individual governments to exercise much tighter control of companies and to favour national concerns in order to ensure that they act in the national interest.

And this is as true of the EEC as it is for the consuming world in general. It is possible to foresee the EEC Commission acting as the general co-ordinating body for national bilateral deals, gathering information, supervising the oil companies and organizing broad discussions and possibly negotiations with the producers on behalf of Europe as a whole.

But whether the Commission will achieve its ideal of ensuring

* In the wake of the 1973–74 crisis several governments announced plans to cut down oil consumption in the market – Holland, by increasing the use of natural gas to a target of some 65 per cent of its energy consumption, and France with an even more ambitious programme to restrain oil imports at their 1973 level, largely by curbing drastically the oil heating market as well as investing on a massive crash programme for the development of nuclear energy. Whether these programmes will survive a relaxation in oil tensions is uncertain.

equal sharing of energy sources, harmonization of prices and federal control of investment and energy development is a very different question. The crisis of supply cut-offs and embargoes found the European nations deeply divided amongst themselves on how they should respond and the experience has only served to divide them further. The discoveries in the North Sea have acted to sharpen the conflicts between national and international development of resources. The differing relations of individual countries with the Arab states, the variety of relative strengths and weaknesses in indigenous energy resources and oil imports and the political divisions on questions of the Atlantic alliance and foreign policy have all served to move Europe in a confederal rather than federal direction on energy amongst other questions. The whole basis of the Commission's role in European energy may thus become the subject of radical re-examination in the future.

Ultimately, however, the really important consumer decisions for the future may be less their developing policies on oil than their action to develop other forms of energy. It may be that there is a final unity of interests between the producers and the consumers that will enable oil to flow freely once more. It may be that price alone will solve the problem by bringing forth new supplies and adjusting demand to create a reasonable balance on oil supplies. But it may equally be that oil remains in tight supply, that the interests of the producers and consumers are in the end contradictory, that the producers force confrontation by attempting to raise prices again and that the oil weapon is once more drawn out of its scabbard. In this case, it is the consuming governments' determination and investment in their areas of direct interest – coal, nuclear power, gas and synthetic fuels – which will ultimately determine whether the consumer is left as naked and afraid at the end of the decade as he has been at the beginning.

CHAPTER 24

Big brother

'It can be summed up in one word that best characterizes this nation and its essential nature. That word is "independence".'

*President Nixon**

In the jostling to form new relationships between the producers and consumers, to develop alternative energy sources and to ensure a better supply–demand balance in oil over the next ten years, it will be the United States which will probably play the most influential role. Its continued dominance of world energy affairs may not be entirely welcome to the rest of the world, indeed it never has been. But its paramount importance on the scene can hardly be doubted.

This is not only because United States' companies still form by far the largest single element in oil production, oil refining and oil marketing in the so-called Free World, accounting for over 62 per cent of the petroleum industry's gross investment in fixed assets at the end of 1972 and around 56 per cent of the industry's worldwide investment in the same year.† Nor is it only because the United States forms by far the largest single national producer and consumer of oil and energy in the world, accounting for over a quarter of world oil production and about 30 per cent of world oil consumption as well as supplying much

---

* Address to the nation on energy policy, 25 November 1973.

† The proportion of worldwide capital expenditure handled by United States' companies has in fact been declining over recent years, while in terms of the gross investment of the industry outside the United States their share is relatively smaller at around 33 to 40 per cent in most areas. The figures are from the admirable series of studies of the industry's finances published by the Chase Manhattan Bank every year.

of the technology, expertise and skilled manpower of the petroleum industry. Nor is it simply because the United States continues to be the world's greatest military and commercial power, and as such cannot be ignored by other parts of the industrialized West, by the consumers or by the developing world.

If the importance of the United States on the world energy scene rested simply on these accounts, one might expect it to decline over the coming years. The 1960s have seen Western Europe emerge as an area of major industrial and political importance in its own right, and, on pre-crisis predictions at least, its position as an oil and energy consumer might well approach that of the United States by 1980. This emerging power is bound to be reflected in a desire for greater political independence of the United States, an increasing squeeze on United States' investment and predominance in oil trade and an attempt to develop technologies for nuclear development and for oil production and transportation in competition with those in North America. And the same kind of spirit of *la defie Americaine* is also partly true of Japan and the developing world, where governments have increasingly moved either directly to take over American assets or to ensure that their predominance is reduced by greater development of national companies.

Even on the broader strategic and political questions, the influence of the United States is beginning to recede in face of renewed pressures towards isolationism at home and growing frictions with its erstwhile allies abroad, although the decline is nowhere near as fast or as radical as the critics of United States influence would like to believe. The Arab–Israeli war of 1973 and its aftermath of embargoes against the United States led by her most trusted friend, Saudi Arabia, were certainly greeted by some as the beginning of the end of American predominance in the Middle East. And the events of 1973–4 will undoubtedly open the door even wider for the Japanese and European investment and trade negotiations that are already growing fast not only in the 'radical' anti-American countries like Libya and Iraq but also in the previous American preserves of Saudi Arabia and Iran.

But United States' foreign policy has never been as slavishly devoted to the defence of United States' oil investment abroad, for all their importance to the American economy and balance of payments, as critics have suggested. Again and again when it has come to a crisis over the oil companies, as it has done at various times in Venezuela and Japan as well as the Middle East, the American government has acted as a moderating influence on its oil groups. And, from this point of view at least, the increasing take-over of producing assets in the Middle East and the cathartic experience of the Israeli War and its aftermath is likely to open up a new and more realistic era of United States–Middle East relations in which America continues to maintain a surprisingly strong presence and influence because of her peace-keeping role.

United States' government's plans for energy independence

| | (*million barrels a day of oil equivalent*) | | | |
	1973	*1976*	*1980*	*1985*
Oil	10·9	11·6	14·0	15·3
Shale	—	—	0·5	1·5
Natural gas	11·2	11·5	13·2	15·0
Coal	6·9	8·4	11·0	12·1
Hydro	1·4	1·4	1·5	1·6
Nuclear	0·1	0·4	1·3	2·6
Geothermal	—	—	0·6	1·0
Total indigenous supply	30·5	33·3	42·1	49·6
Demand at reduced growth rate of 2% per annum	36·6	38·8	42·1	46·5
Net imports required	6·1	5·5	0	(−3·1)

Source: *Oil and Gas Journal*, March 1974

Yet the really fundamental importance of the United States on the world's energy scene over the future may well lie much less in this kind of political and commercial activity than in its own position as a major oil importer in its own right. After nearly fifty years of fulfilling the role of a balancing wheel in international oil, the 1950s and 1960s saw America withdraw more and more into an isolated position in the world's oil trade, depending for its imports on Western Hemisphere sources like Venezuela, Canada and Mexico and surrounding itself with tariff protection against imports. Now it has emerged as a potential market for the Middle East and Eastern Hemisphere on a really daunting scale. Indigenous production of United States' natural gas passed its peak nearly five years ago. Oil output started a gradual decline, without the benefit of major Alaskan flows, at the beginning of the 1970s and the growth in demand for energy has had more and more to be supplied from imports and, more recently, from Middle East and African supplies above all.

By the time of the Arab oil embargoes against the United States, the country was already dependent on imported oil for about a third of its total oil consumption, was increasing its imports at a rate of some 30 per cent a year and was calling on Middle East oil to the extent of nearly 2m. barrels per day, including oil shipped to the Caribbean for refining and then imported as products. Given a continuation of these trends, all the predictions suggested that by the early 1980s the United States could be importing half or more of its total oil requirements at some 10–15m. barrels per day, the equivalent of the entire Middle Eastern output in 1970 of which two-thirds would have to come from Africa and the Gulf.*

It is America's ability to withdraw from this situation which, more than any other single factor, is likely to influence the course of prices and supplies to the rest of the world over this decade and possibly well into the 1980s as well. If it does continue to increase its imports at previous rates, then the weight of its purchasing power and the volume of the needs could strain the oil trade to breaking point and leave Western Europe and Japan in serious difficulties in any competitive rush for available supplies. If it can work towards greater self-sufficiency over the medium term

---

* The suddenness with which this situation has hit America can be seen from the fact that as late as 1970 the President's Task Force on Oil Imports, taking its figures from government and industry, was predicting that demand in the United States would not reach 18·5m. barrels per day until 1980 and that imports at that time would total 5m. barrels per day, most of it coming from the Western Hemisphere. In fact the demand figure had already been reached by the time of the Arab embargo in 1973, at which time the United States was importing over 7m. barrels per day, over 2m. of which were from outside the Western Hemisphere. The miscalculation is almost a re-run of the situation in Europe in 1970–71 when industry and governments alike found their estimates of demand and imports overtaken completely.

and a drastic reduction in the rate of growth in its imports over the short-term, then the possibilities of oil supplies in the rest of the world becoming easier would be much higher. Either way, it will be extremely difficult for the rest of the world to produce any kind of secure arrangements for oil supplies over the future, or to develop alternative energy sources, without reference to the United States' own demand for imported oil and its own rate of development and research into other forms of fuels.

The reasons why a country with more natural resources than virtually any other area of the world should get into this situation, and suffer acute local shortages of heating oils, gasoline, fuel oil and natural gas over the period 1972–74, have become the subject of passionate debate in the United States. For some, the whole problem has been simply the result of environmental pressure from 'ecofreaks' who have managed to put an almost total stop to the construction of new refinery, distribution, terminal and pipeline facilities that were necessary to sustain the rate of growth in America. For others, it has been the oil companies which have created the situation purely in order to raise prices and do away with the competition from independents. For some it is geology that is at the back of it, for others it is not geology but government policy and the absence of real incentives that has produced the 'crisis'. Some even doubt whether there is a crisis at all, while other commentators see in the situation a reflection of a deep-seated crisis of a way of life that has put consumer consumption and constant growth above any consideration of the finite nature of resources or the damaging effects of pollution.

It would be trite to say that it was all these things, although it would not necessarily be inaccurate. American public debate has always suffered less from the absence of hard information than a plethora of it, and there are 'facts', or at any rate assertions, which could support almost any view on the causes of America's current energy problems.

The most straightforward explanation, of course, is simply the fact that unceasing growth in demand has finally begun to strain easily-available supplies. For longer than almost any other country, the United States has been a major consumer of hydrocarbon fuels, with an energy consumption rate per head of population double that of Western Europe and Japan and nearly eight times that of the developing world. Driven by a consumer-orientated society seeking ever higher standards of personal comfort and mobility through the acquisition of heating and air-conditioning processes, labour-saving devices and bigger and more powerful cars – all of which entail a high consumption of energy – oil demand has reached the stage where it is growing by over 1m. barrels per day each year – the equivalent of the entire oil output of Algeria. Put another way, the United States now demands the discovery of a new North Slope of Alaska every two years merely to keep pace with demand. And, while some very large oil provinces have been found in Texas and elsewhere, the country has never had the kind of vast, easily producible, low-cost fields of the Middle East. Over the last five years and more the amount of oil discovered per foot of drilling has declined. Outside of Alaska, new finds are often small and the ratio of reserves to production has been falling for several years and has now reached only one in ten (enough to last ten years at current rates of output), which is probably the minimum level which can be safely sustained.

Where United States' energy goes

Market	%
Industrial	32
Electrical generation	25
Transportation	24
Residential	14
Commercial	5

Exactly the same thing has occurred with natural gas output, which accounts for over a third of overall energy consumption in America. Fed largely by artificially low prices, demand for gas in the United States has been galloping forward at a rate of 6·5 per cent a year, requiring an additional 1,000m. cubic feet per day level of supply each year and implying a doubling of demand by 1985.

But, once again, the rate of new gas discoveries, although for different reasons than for oil, has failed to keep pace with the rise in demand. Proven reserves have fallen to the equivalent of around eleven years' supply at current rates of production, compared with more than sixteen years' supply only a decade ago, and there are strong grounds for believing that, unless new reserves can be found and developed quickly or a massive programme for the manufacture of substitute natural gas (SNG) from coal can be undertaken immediately, indigenous gas supplies in the United States will decline by as much as 33 per cent over the next ten years against a potential demand that could rise by more than 50 per cent over the same period.

On this situation of stagnating oil and gas production, the environmental pressures which have come to a head over the last few years have had a devastating impact. Spawned both by specific incidents of oil pollution from accidents, and by a general feeling of unease over the possible cost in ecological terms of unrestrained industrial growth, popular protests had managed to bring a virtual halt to off-shore oil production on the west coast, to delay by nearly five years the construction of the Alyeska pipeline necessary to bring the giant North Slope discoveries in Alaska into continuous production and to slow down the rate of development not only of new gas and oil reserves, but also the construction of new refinery plant, power stations as well as terminal and distribution facilities for large-scale imports of oil and gas. It would be too easy to ascribe all America's current problems of shortages to these factors. But undoubtedly environmental pressures have helped very considerably to intensify a situation already headed for severe trouble and coming at the time that they have, to increase greatly the difficulties of adjustment to new sources of fuel.

Nor can successive United States' government policies over the last few decades be entirely absolved of the blame for some of America's current energy confusions. More than almost any other industry in the United States, oil and gas development have been afforded a unique degree of government protection, price interference and tax support at the same time as displaying some of the more riotous aspects of unrestrained competition and a near-religious devotion to the rights of individual property ownership when applied to the subsoil.

Although this has inevitably raised foreign eyebrows at the hypocrisies involved in a country preaching the absolute virtue of *laissez-faire* economics abroad while busily protecting its own industry at home, the development of government policies in the United States has probably been less the result of blatant nationalism than the specific circumstances of the history of its oil industry and the power which the major oil and gas producing states such as Texas have held in local and national politics. As early as 1913 the government supported the search for oil and gas with the famous, or infamous, depletion allowances, under which oil and gas producers can set off a portion of their income against tax for the progressive loss of their basic assets (oil and gas) through production.

The theory of this tax is logical enough – that the producer of oil and gas is in the unique situation of not adding to the value of his assets with time but actually reducing it as the oil and gas is depleted and that, therefore, he should be entitled to some recompense for the loss. But the allowance, eventually rationalized at 27·5 per cent of the price of the fuel before being reduced to 22 per cent in 1970, has given the oil, as well as other extractive industries, a substantial benefit. And it is one that has grown more and more advantageous with time as the rate of corporation tax which it relieves has risen. And to this other tax advantages have been added, most notably the right to expense rather than capitalize most of the investment associated with the discovery of oil and gas, including 'intangible costs' – a treatment that allows oil companies immediately to write off the costs of exploration and development against income rather than having to capitalize the costs and write them off over a number of years as in the case of most other businesses; the right to offset directly against United States' tax liability on foreign income, tax payments made to foreign governments – a measure to prevent 'double taxation' which has become progressively more valuable as the producer government 'take' has expanded; and finally the right to claim depletion allowances on foreign production.

Even more valuable to the oil- and gas-producing industry in the United States over the last decades has been the 'pro-rationing' systems which most of the major oil and gas states have adopted to rationalize the overall rate of output in their states. Here again, there have been perfectly logical reasons for this. American rights on private property and the 'law of capture' (i.e. oil belongs to he who can produce it), resulted during the years between the wars in a destructive proliferation of wells on every find as rights to produce were sold off in small lots and each owner desperately strove to produce as much as possible from his piece of land. If for no other reason, therefore, a pro-rationing system which could look at the field as a whole and ensure that the wells upon it were controlled so as not to deplete the field too fast and upset its producing characteristics became a necessary form of official intervention.

But once set in motion, the business of controlling output of fields has all too easily grown into a price protection system. Pro-rationing in the name of efficient production quickly became pro-rationing according to market demand in a number of important states, especially in Texas where the Texas Railroad Commission for years sat each month to decide the output of the majority of wells in the state on the basis of demand forecasts presented by the industry and government officials. And the effort of individual states in this direction was backed by the federal government through the Connally Hot Oil Act of 1935, which was introduced to prevent the interstate movement of oil produced in excess of state allowables; and the successively-renewed Interstate Compact to Conserve Oil and Gas of the same year, which established mechanisms for co-ordinating the rationing schemes of individual states against overall national estimates produced by the Bureau of Mines.

Once rationing against demand became established, then the measures inevitably worked towards maintaining price structures by ensuring there was no surplus supply to undercut prices. And once this had been done the third, and perhaps most dramatic, of all indigenous oil-support measures – protection against foreign imports – became equally inevitable. Almost as soon as the Second World War was over, domestic production came under threat from the growing competition of lower-cost oil from Venezuela, the Middle East and other parts. By the mid-1950s the threat was overwhelming with the expansion of output from the Gulf and, even more, the discoveries of some of the United States' independents moving into North Africa and the Middle East. By 1958, imports, which had exceeded exports for the first time in 1948, had captured some 18 per cent of the United States' domestic market, and the Texas Railroad Commission was restricting the output of controlled wells in the state to a mere eight days per month.

Successive failures to adopt a voluntary system of restraint on imports finally forced President Eisenhower to act in 1959 to introduce mandatory quotas on imports, restricting imports of oil to around 12·2 per cent of demand in the major consuming area east of the Rockies allowing imports to fill the gap between production and demand in the western region and giving preference within the quotas to Canada and Mexico. Thus was begun a system which effectively isolated United States' prices and supplies from the Eastern Hemisphere for more than twelve years, forced United States' producers to find an outlet for their oil in Europe and Japan and firmly buttressed the pro-rationing system of the States of America.

It is this trilogy of tax support, production controls and import restriction which has given the United States' oil industry so much of its peculiar flavour of vast fortunes being made by a few individuals at the drop of a drilling bit, of armies of lawyers devoted solely to the problems and possibility of oil claims, of endless streams of money in search of tax avoidance and of companies of immense size jostling side by side with the penniless wildcatters in search of an overnight bonanza. It has all the elements of the Hollywood dream of success to the poor boy who takes a chance coupled with some of the worst aspects of corporate greed, political machination and government manipulation.

Whether it has been to America's benefit or not is a rather different question. Certainly it has enabled the United States to continue with a surprising degree of self-sufficiency of energy over the last generation or so. As much as 2–3m. barrels per day may have been produced that would not have been under foreign competition, and although this protection has been costly – even government estimates have put it as high as $4,000–5,000m. a year in the form of higher prices towards the end of the decade – it might well be argued that recent events in the Middle East and the Arab embargo on shipments have shown it to be worthwhile.

But the policies have also had some more debiliating influences as well. Although accusations of oil company tax avoidance often ignore the high degree of local state taxes that companies pay, there must be strong arguments that the price and tax benefits afforded to the oil industry have been greater than has really been needed to achieve their purpose of buoyant indigenous production and that the tax benefits enjoyed by the industry have been less productive than they should, perhaps, have been.*

Of more pernicious impact has been government intervention in the fixing of gas prices in the United States, although gas production has never been subject to the same pro-rationing controls as oil, simply because it has not the same flexibility of delivery and, when associated with oil, has to be produced at a rate set by the oil output. But, since 1938, the transmission companies buying gas from the producer and transporting it to markets in other states have been subject to the overall control of the Federal Power Commission, which also gained control of the prices at which the transmission companies bought gas through a Supreme Court decision of 1954.

However worthy may have been the intentions of the FPC and

---

* The whole question of tax treatment of the oil industry in the United States has aroused an endless stream of charge and counter-charge. Critics have tended to argue not only that the tax treatment has been unfair and unproductive but that it has also increased the tendency of vertically-integrated companies to concentrate their profit centre at the producing end to the detriment of fair competition in the market-place. Defenders have equally claimed that this has not been the case as evidenced by the profit record of various companies and that the tax benefits have not been nearly as much a bonanza for the industry as opponents have claimed. Into this argument only the most foolhardy of foreign observers would stray. But it is probably fair to say that, considering the political anger that they have aroused and the high prices for domestic oil that have come into being, they will probably be dropped in the near future and may not be missed. What appears to be replacing them is a combination of price controls on domestic oil production, tax incentives for energy investments and, possibly, an excess profits tax to cream off politically unacceptable returns.

13. Super-Tankers: (1) The launching of *Esso Mercia* (170,800 tons)

14. Super-Tankers: (a) Loading at Kuwait

the Supreme Court, the decision to control gas prices effectively placed a burden of artificial calculation of costs and returns on production coupled with a sale-by-sale procedure for determining the price of each contract that the Power Commission has been unable to cope with. Gas prices have been unduly depressed as the Commission has got more and more behind with its cases and the issue has entered the realms of politics, where no official would lightly take on the responsibility for approving higher prices. Exploration for new gas reserves has meanwhile fallen dramatically while demand, geared by low prices, has shot ahead at a faster pace than any other fuel and has taken the largest share of most industrial fuel markets. In almost the reverse situation to oil, the United States' consumer has benefited enormously from two decades of low gas prices while the overall national interest in securing expanding gas development has suffered correspondingly.

Added together, gas price regulation, environmental pressure against various developments, state intervention in oil production and declining indigenous reserves of both oil and gas have all served to put the United States for the first time in a situation of actual shortages, electricity cuts, mandatory allocation systems by the government and severe local disruption. In a situation where shortage of refining capacity was already looming on the horizon by the beginning of the decade, government indecision about whether to drop the quota system or not, its introduction of oil price controls in 1972 and environmental objections to the siting of new plants have all acted to worsen greatly the situation and produce two years of uncertainty about gasoline, heating and fuel oil supply.

Under conditions where production of indigenous oil and gas were already flagging, the delays in getting the Alyeska pipeline underway have only made the country even more vulnerable to the kind of interruption in foreign supplies which the Arabs introduced over the period November 1973 to March 1974. In circumstances where artificially low gas prices were already priming gas demand, controls over the use of high-sulphur oil and coal have only acted to raise gas demand further as the most pollution-free fuel just as regulations to reduce the lead content in petrol have tended to increase the amount of petrol required.

A country which more than any other has taken unlimited supplies of fuel as a matter of right was thus faced in 1973/4 with the traumatic experience of having to queue for its petrol, accept electricity cuts almost as a seasonal occurrence and have its supplies of propane, butane and other products allocated by the government according to national priorities. No such interference has ever been introduced outside of wartime and, although by the summer of 1974 the United States seemed to have scraped through the worst of the Arab embargo without as much trouble as had been expected, it was still far from over its problems.

The great question now is whether this experience will be enough to provide the much-needed public push to enable the country to achieve the turn-around it is now looking for. Certainly it has the technical capacity, given the diversion of resources into this field, and certainly it has the potential reserves with its vast resources of coal, shale oil, uranium and conventional hydro-carbons. Estimates of oil and gas still to be discovered vary widely, but at the minimum they suggest that the country may still have as much as four times its current level of gas and oil reserves still to be discovered.

On top of this, the country has unequalled opportunities for rapid development of its immense reserves of coal, estimated to total some 500,000m. tons, or enough to supply the country for several hundred years at current rates of demand, and is now pursuing a very promising line of research into the manufacture of substitute gas or oil from coal. Recoverable oil from shale oil may amount to as much as 80,000m. barrels of oil, or double the present reserves of conventional oil, and here again research is now being dramatically expanded into ways to produce the oil either through *in situ* methods or by mining and then extracting the oil. Programmes for research and development of nuclear power are being revised at a rapid rate and it is in the United States as well that one might expect the greatest advances to be made into the development of solar and geothermal power.

The United States, at the same time, may also hold relatively greater opportunities for savings on fuel use than other nations, in view of its very high energy consumption per head and the relative inefficiency with which it converts energy to useful power in its large motor cars, its unusually luxurious standards of heating and cooling and its relatively high consumption way of life. Government action during the crisis to regulate the levels of heating in offices and factories, to allocate supplies of certain fuels and to encourage greater conservation of energy have proved that fairly dramatic reductions of as much as 10 per cent and more can be achieved without great damage to the economy while the move towards smaller cars with lower petrol consumption, which are already going on, could gradually reduce the level of increase in petrol consumption to a point where it is no greater by the mid-1980s than it is today, if, say, 60 per cent or more of cars on the road were then 'small ones'. And motor gasoline alone accounts for nearly half of United States' oil consumption.

All this could mean a dramatic move away from import dependence within a relatively short-time scale of this decade, certainly faster than any other country could hope for. The North Slope reserves could be brought in at around 2m. barrels per day by 1977–78 and possibly reach double that figure by the early 1980s, while the substantial increase in exploration activity, already being seen as a result of high prices and greater profitability on the part of the companies, should yield results in terms of both greater gas and oil production by around the same time. Coal production could be stepped up by as much as 60 per cent or more over the next seven to ten years if a major investment in strip-mining takes place. Immediate problems could be relieved by continued regulations to dampen use, by relief on some of the regulations controlling the burning of sulphurous fuels, by the opening up of military reserves of oil and by greater recovery from existing fields now made economic by the higher prices. And to back this effort up, the United States' government has now instituted an energy programme designed to increase Federal funding of energy research to $10,000m. over the next five years; to open up federal lands and off-shore acreage to more rapid exploration as well as shale oil concessions; to streamline government supervision through the creation of a centralized Federal Energy Office; to ease up substantially, if not completely, on the regulation of gas prices and to relieve some of the current pressures created by environmental legislation.

But there remain difficulties to any short-term import sub-stitution. There are formidable environmental objections to both the present generation of nuclear power stations and to rapid development of coal through strip-mining. The problems of sulphur in coal may not be as easily overcome as some in govern-ment might think, and may indeed have to await the development of large-scale units to produce clear synthetic natural gas before coal can really come into its own.

The development of off-shore reserves, the build-up of plants to convert shale into usable liquid oil, the move towards smaller cars, meanwhile, will all take at least five years before they can make a large-scale contribution to the United States' energy scene.* Prospects for energy conservation, too, may be easier described than instituted in view of the large proportion of the United States' population living in small towns and completely dependent on the car for any real transport and in view simply of the difficulties in a democratic society of forcing people to change their individual habits other than through response to prices.

Finally, and not least, there is the question of where Govern-ment policy itself will stand on questions like the environment, import control and price and profit regulation. If the problems of the early 1970s have finally awoken United States' public opinion to the real energy difficulties that the country is now facing, they have also aroused a tempest of accusations against the oil companies for deliberately manipulating the difficulties to their own advantage and left widespread confusion in the public mind over just how deep-rooted the crisis is. Under these con-ditions, the Government, as the Arab embargo was finally eased in March of 1974, was still moving uncertainly on questions of whether to remove entirely import regulations or whether to find some other mechanism to control them, whether to introduce

---

* Pre-crisis estimates by the National Petroleum Council of the United States suggested that by 1985 production from oil sands might amount to between 500,000 and 1·25m. barrels per day; with a further 100,000–400,000 barrels per day coming from oil shale and perhaps 100,000 barrels per day of synthetic oil being produced from coal. These estimates could prove conservative in the light of what has happened since, although they are unlikely to be too much out.

new profit taxes on the companies or whether to control prices instead, and whether to totally de-regulate gas prices or simply raise the ceilings, while the old talk of forcing the oil companies to break up their vertical integration in refining and marketing was once more growing.

America's answers on these questions, as on others, could have a profound influence on not only its own industry but conditions elsewhere. If the United States does continue to depend on increasing imports over the coming years, particularly supplies from the Middle East, then the pull on Eastern Hemisphere supplies and prices could be very strong indeed and the position of United States' companies abroad could take on a very different shape.

This shape could also be changed if the government – as it well might – moves to reduce some of the tax benefits such as depletion allowances for foreign drilling and tax credits on foreign earnings that have so long encouraged United States' companies to expand and drill abroad. Under these kinds of circumstances, some of the United States' oil companies might tend to confine themselves to much more of a trading role abroad, concentrating on buying oil for the domestic market in the Eastern Hemisphere and concentrating more and more of their investment in drilling, refining and marketing back at home, where the profit opportunities might be greater. Already some of the major companies with less extended foreign interests, like Gulf and Socal, have been reviewing their foreign investments in a searching light and there are signs that a restoration of investment balance on the part of the United States' major companies, which has tended to favour foreign investment for many years, could now swing back in the other direction.

But against this are the possibilities that the United States' government, pushed by public antipathy to the oil companies, might act to cream off much of the expected profit opportunities now presented by higher prices and tightness in supply; could continue to regulate prices to the detriment of development; could abandon the whole system of tax protection and could even encourage the disinvestment of much of the industry's assets in refining and marketing on grounds of anti-trust or fair trading.

In this case, not only will the country's prospects for moving away from import dependence be changed, but the incentive may be created for United States' companies to decentralize and increase their investments abroad where profit opportunities may still be good.

Suggestions that the United States' government is deliberately exaggerating the degree of difficulty now facing the country on energy may well be right in the sense that these difficulties are less the product of a real crisis than a temporary dislocation in supplies caused by government regulation of gas prices and environmental objections to the construction of new refineries and the development of new reserves. But the problems do reflect what might happen if nothing is done about the situation.

It is the United States' long-term response to the 'crisis' that matters to the rest of the world, not the rights and wrongs of how it came about. No one who has ever visited the country can doubt its capacity, given the will, to find a solution to its energy difficulties. But no one who has ever visited the country can equally doubt its capacity to avoid and confuse action in a welter of political arguments and irrelevant obsessions with past misdeeds. Even with the fullest possible effort it is doubtful whether the country could really become completely self-sufficient in energy within this decade. But without a very substantial effort it could become more and more dependent on foreign imports on a scale that would have profound consequences on the world's oil trade as a whole.

CHAPTER 25

'The business in between'

'The companies, as private organizations and under
the terms of reference applicable to commercial
corporations, cannot possibly be expected to carry by
themselves the burden of protecting not only their own
interests but also those of their customers.'

*Walter J. Levy**

The most obvious casualty, or culprit as some may prefer to put
it, of the prolonged series of oil crises over the last few years has
been the industry itself, and the major international oil companies
in particular. As with most middlemen in a period of commodity
scarcity, the oil companies have made considerable profits out
of the troubles of the new decade and, in so doing, have earned
even greater showers of public opprobium. And in their case, the
damnation has been made all the more virulent by their own
history, their size and the myths of supra-national power that
have always surrounded them.

Almost nothing they have done in the past, or are likely to do
over the future, can make them popular, or the virtuous reflections
of responsible business practice that they sometimes tend to see
themselves as. At a time when big business has become almost a
dirty word in the language of popular politics, the oil industry
is not only big but *the biggest* business in the world. At a time when
economic nationalism has become the driving force of almost
every developing nation state, the basic internationalism of the
oil companies is seen not as a benefit but as a threat, the product
of rampant capitalism or the tool of neo-colonial domination by
the great powers.

If they fail to make reasonable returns on capitalism, they are

* In a speech on 'An Atlantic–Japanese Energy Policy', March 1973.

accused of being like dinosaurs, failing to adapt themselves to a different world. If they do make money, they are criticized for manipulating the woes of others, feeding on their opposition and creating crises for their own profit. If they attempt to share out supplies between all their customers at an international level, they fall foul of each individual nation state which feels that it deserves favoured treatment. But if they do not share out supplies equably, they are accused of moving the oil round to where the prices and profits are greatest or of serving their particular government masters in Britain, the United States or France.

And yet, the dramas of the last few years, far from displaying the almost imperial-like power of which the oil companies are so often accused, have in fact shown up how little real power they have now come to possess. The industry–OPEC negotiations of price at Tehran and Tripoli, and later in Geneva, almost certainly marked the last occasion on which the oil companies attempted to negotiate on their own a price structure for the oil trade. This was one of their central roles during the 1960s. But it was only operable when supplies were in surplus and a genuine balance of power existed between the producers with their oil and the oil companies with the outlets for that oil. Once that balance was upset by tightening supplies and the industry was revealed to have no real flexibility as to supply sources left, then OPEC was bound to brush them aside as negotiating agents for prices – as it did in October 1973. Far from being able to resist or manipulate the producers, the oil companies are now only able to accept their price demands and pass them on to the consumer to see whether they are prepared to fight them or not.

The same could be said of the field of supply allocation, where the industry's efforts to share out available supplies over the winter of 1973–74, however successful, may well prove to be the last occasion on which they take this burden on their own back. It was their ability to redistribute oil successfully during the two closures of the Suez Canal and the embargoes and supply disruptions which accompanied them in 1956 and 1967 that made the major oil companies such valuable middlemen between producers and consumers in previous decades. On the last occasion, not only were the companies unable to do this as effectively as they had done in the palmier days of the past but their efforts all too often aroused the ire of the consuming governments. Countries like France and, to a lesser extent, Britain, who had been assured of special treatment by the Arab producers, baulked furiously at the idea that some of their non-Arab supplies should be diverted to countries like Holland most directly hit by the embargoes.

Regional balance of the major oil companies

	1962	1968	1972
Gross Production (m. Barrels per day)			
Western Hemisphere	1·24	1·52	1·49
Eastern Hemisphere	2·31	4·39	7·06
Net earnings ($m.)			
Western Hemisphere	588	843	769
Eastern Hemisphere	1,227	1,782	1,995
Net assets ($m.)			
Western Hemisphere	5,078	5,542	7,210
Eastern Hemisphere	9,353	14,245	19,882
Return on net assets (%)			
Western Hemisphere	11·6	15·2	10·7
Eastern Hemisphere	13·1	12·5	10·0
Earnings per barrel produced (cents)			
Western Hemisphere	47·3	55·6	51·5
Eastern Hemisphere	53·1	40·6	28·3

Source: A study by the First National City Bank of New York into the operations of BP, Exxon, Gulf, Mobil, Shell, Standard Oil of California and Texaco.

Other states, like Japan, who were caught in the middle of the Arab policy of discretionary treatment of their customers, felt just as keenly that the oil companies were not doing enough to help them. And the task of the oil companies was made only more difficult by the differing degrees to which individual companies depended on Arab production for their supplies and the rash of restrictions imposed by national government on the export of products from their refineries. Even though the allocation effort by the companies was reasonably successful, partly because the Arab intervention into destinations and quantities of exports was never as stringent as it appeared on the surface, it is unlikely that the oil companies will ever take on the full responsibility of repeating the experience without the full support and clear agreement of all the consuming governments.

And if the oil companies have clearly lost a great deal of their traditional ability to bargain with the producers over crude oil prices and effectively distribute supplies, there seems little evidence to support the belief that they have the power to fix prices in the markets either – despite the accusations that the 1973–74 crisis was an artificial one created by the oil companies in order to raise price and profit levels for oil products. Certainly it has not been against the commercial interests of the oil companies to see the tightening supplies, falling competition and rising prices that have accompanied the oil problems of the last few years. But the reductions in Middle East output were announced not by the oil companies but by the Arab producers, and if they were never quite as tough as they were said to be, a great deal of extra oil slipping by came through the hands of independent operators and the state oil companies rather than the international companies. The panic in the ranks of consumers and oil companies alike was based on the psychological expectation of cold weather and further action by the Arabs, both of which were credible enough predictions at the time in view of Arab statements and the law of averages on weather even if they did not in fact come to pass.

Far from being a product of deliberate oil company machinations, indeed, the rapid rise in prices and the consequential increase in profit levels of the major groups in 1973 could more simply be explained by the state of competitiveness that had been built into the oil trade over the previous decade. In any situation of real competition, prices tend to follow what is happening in the margin, in other words the lowest costs at which any individual company can obtain oil and sell it at a reasonable profit, although the amount of oil sold on the margin might total no more than 5–10 per cent of total sales in most countries.

During the 1960s, when crude oil production, tanker carrying capacity and refinery capacity were all in surplus, the marginal price at which individual traders could gain supplies was low, prices fell and the major oil companies, with their higher average costs, suffered. Once either tanker carrying capacity, as in 1967–68 and 1970–71, or crude oil availability, as in 1973, became tight, then the cost at which independent traders could gain marginal supplies went up, prices in the market followed and the oil companies, with what then became lower average costs, benefited.

And this logic of free market movements does much to explain what happened in 1973, when, long before the Arab cut-backs, tightening crude oil supplies coupled with a general expectation in the market of a difficult winter ahead (made more tense by potential United States' demand for product imports) tended to

Crude oil production by the majors in 1972
in thousand barrels per day

	West Europe	Far East	Africa	Middle East	Latin America	North America	Total	% of Free World production
Exxon	59	190	307	2,283	1,519	1,376	5,734	13·0
BP	26	1	603	3,995	6	18	4,649	10·6
Shell	39	292	760	1,649	945	819	4,504	10·2
Texaco	31	429	122	2,100	275	1,603	4,020	9·1
Socal	1	427	122	2,078	63	592	3,283	7·5
Gulf	4	—	452	1,881	250	688	3,275	7·4
Mobil	22	36	272	1,135	140	589	2,194	5·0
CFP	—	—	151	895	—	5	1,051	2·4
Total majors' production	182	1,375	2,789	16,016	3,198	5,150	28,710	
Total production	332	1,882	5,526	18,337	4,913	13,046	44,036	
Percentage held by majors of total production	54	73	50	87	65	39	65	

raise the price of oil on the marginal markets, reduce the surplus available to meet rising demand and intensify the pressure of consumption against available supply. The major oil companies were unable to take as full advantage of this as the independent traders – who could stock oil against a further rise in prices and switch it from low-priced to high-priced markets – because of the price control exercised by many Western governments, including the United States, and the long-term contract obligations of the international oil companies to their customers. But they were able to reduce some of the sizeable discounts offered to bulk buyers that had been one of the strongest features of marketing during the 1960s, to raise the level of returns in some of the lowest-priced areas such as aviation and bunker fuel and to shed some of their unprofitable business. And it was this process of reducing commitments and cutting back on discounts rather than raising official price levels which was more than anything responsible for the upsurge in profits for the year.

The process was hardly welcome to those customers who had gained the greatest benefits from the low marginal prices of the previous decade – the 'spot' buyers of oils on the open market and the 'ex-tank' buyers of surplus products from the major oil companies. But rising prices and increasing profits for commodity traders was something that occurred in almost every industry experiencing tight supply conditions in the early 1970s, from steel to man-made fibres or natural wool, and requires, in oil's case, no special explanation in terms of the structure of the industry or the dominance of the major companies in its trade.

The aura of immense supra-national power held by a few major oil companies has in fact reflected reality less and less closely over the past decade or so, if indeed it was ever true in the sense that some critics have claimed. The international oil companies may well control budgets that are several times as large as most African, Asian or Latin American states. Their freedom to invest and switch funds around their systems has certainly produced real conflicts between their corporate interests and the national interests of the countries in which they operate. But the fact that their investment is in the form of physical asset such as refining, marketing and production facilities has in many ways made them more vulnerable to nationalist pressure than if they had been purely traders. As the nationalization of those assets in innumerable small countries has shown repeatedly over the past, there is little an oil company can do when its operations come under attack, however small the country involved. And, as the long series of negotiations with governments over prices, state participation and concession terms have shown at both the producing and consuming ends of the business, once the investment in those assets has been made, the natural tendency of companies is to cling on to them however high the cost.

Nor has the industry's size, flexibility or geographical spread prevented its position in the market being steadily eroded by competition from new entrants and the intervention of national oil companies over the past fifteen years. In the area of ownership of refineries, the position held by the seven majors (or 'seven sisters' as Mattei of Italy was wont to call them) fell steadily throughout the non-communist world outside North America from well over 65 per cent to around 58 per cent during the 1960s, while the share of the independents increased from less than 20 per cent to over 25 per cent, and the share of governments rose from around 14 per cent to over 17 per cent, with the drop being seen most sharply in Japan and the Far East where the share of the majors fell from over 75 per cent to less than 33 per cent over the decade. In the area of market sales, meanwhile, the share of the majors in the non-communist world has fallen from over 60 per cent to about 54 per cent, while the share held by the independents has risen from over 25 per cent to over 30 per cent and the share of governments has climbed to around 15 per cent.

This is not to suggest that the international oil companies are powerless or that their economic interests are necessarily those of the consumers in some perfect fashion. The primary motivation of the oil industry has been clearly a commercial one of gaining the maximum possible return on their investments consistent with their own best interests of self-survival. And the really extraordinary thing about their development over the last decade or so may lie less in the fact that their position has been eroded than that they have managed to retain quite such a dominating share of the world's oil trade.

At a time when world consumption of oil has doubled over the last decade and product demand in the Free World has risen at a rate of nearly 10 per cent per annum in the Eastern Hemisphere, the majors' continued share of over 50 per cent of the Free World's refinery capacity and over 60 per cent of its product sales still leaves it in a commanding position. Far more important, the majors have been able to own and ship by far the greatest proportion of the Free World's oil production, thanks largely to the growing importance of the Middle East. Their total share of Free World oil output at the beginning of the 1970s, although it had fallen from some 80 per cent at the beginning of the 1960s through nationalizations and the new discoveries of the independents, was still some 70 per cent, while their share of Eastern Hemisphere output was over 75 per cent.

In terms of Middle East output, the share of the majors has been until recently even greater, at nearly 87 per cent of output in 1972 through various consortia such as Aramco (Exxon, Texaco, Standard Oil of California and Mobil) in Saudi Arabia, the BP–Gulf partnership in Kuwait, the Iranian Oil consortium of all the majors in Iran and the various extensions of the Iraq Petroleum Company consortium of BP, Shell, CFP, Mobil and Exxon in Iraq and Abu Dhabi. Despite all the attempts of national oil companies and independents to establish themselves in exploration, the majors retained this dominance of crude oil supply throughout the 1960s and, with the increasing reliance of the world on Middle East supplies for any growth in consumption, actually raised their share of crude oil supplies during the first years of the 1970s.

And it is this dominance of crude oil production, even in the United States where tax incentives have encouraged a much wider flow of exploration investment through the industry, which has enabled the large international oil groups to exercise so much 'muscle' and flexibility in the market as compared to their competitors – more especially as it has been at the crude oil end that the profit centre of the industry has always lain.

The fundamental question now is whether the loss of this production strength through participation and/or nationalization in the Middle East coupled with the intervention of consumer governments in oil questions will lead to a drastic reduction in the position of the major international oil companies over the future or whether, as some have suggested, the loss of ownership of oil will drive them to a more intense effort to dominate the market end. For some, undoubtedly, the recent crises in the Middle East mark the end of the reign of the international companies, as national governments take over the job of negotiating price and supply and begin increasingly to deal on a direct government-to-government basis with the producers. For others, however, there remains the fear that the crises of the last few years have also seen a dramatic reduction in competition in the oil trade and that the major oil companies may actually increase their 'stranglehold' of the trade as a result.

Central though the question is, it is not one for which anybody can predict an easy answer. The role of the major international oil groups is bound to come under increasing pressure over this decade, certainly in terms of their traditional freedom over pricing and supply allocation and very probably in their share of oil trade as well. As the producing countries move towards complete or majority control of their oil output, they will unquestionably seek to sell a proportion of that oil themselves, particularly if market conditions remain in their favour. Equally, the consuming governments, now that they have been dragged directly into the oil question and recognize its fundamental importance to their own economies, will want to influence the way in which that oil is handled. Like the producing governments, they see advantages in direct dealings with producer government to give added insurance to their supplies and, like the producing governments, they recognize that oil is too vital a commodity to leave entirely in the hands of commercially-orientated, international oil companies. At the very least this will mean that the role of the oil companies is reduced in some instances to that of an agent to governments, and the chances are surely that governments will wish to develop their own national oil companies as far as possible over the future.

Yet there remain formidable reasons why a move towards bilateral relations between consumer and producer and an undermining of the role of the international oil companies may not go as far or as fast as some proponents of national oil companies have suggested. One is that both producers and consumers still remain reluctant to commit too much of their oil trade into direct trade deals that might become vulnerable to political conflicts between the parties or rapid changes in economic circumstances.

The producers have shown that they are clearly prepared to use oil for political purposes and the dangers for the consumer of tying up his oil sources too directly to a limited number of crude oil suppliers has been amply illustrated by the experience of the Franco–Algerian agreements and the actions taken more recently by Iraq against United States and Dutch oil assets as part of the Arab–Israeli struggle. Even should the Israeli question recede, there is always the chance that other struggles could erupt in the Middle East – between Iran and Iraq for example – – that could prove even more embarrassing for the consumer should oil become involved.

And, equally for the producers, it may not make too much sense to tie a major portion of exports to particular markets, for commercial as well as political reasons. The flow of goods between the oil exporter and oil importer is never a neat one and both sides may be reluctant to push too far along the line of tying general bilateral trade with oil, although it may make more sense on a regional EEC basis.

A second argument against too dramatic a usurpation of the position of the international oil companies by national oil groups may be simply the scale of oil company investment and the size of expenditure needed to provide for the future. Total oil industry gross investment in fixed assets at the end of 1971 has been calculated at over \$220,000m., of which more than 60 per cent was in the area of transportation, manufacturing plant and marketing facilities. In Western Europe alone, the industry's investment in refining, marketing and distribution amounted to more than \$25,000m., of which more than half was held by the international oil companies, whose net worth in the Eastern Hemisphere in 1972 amounted to nearly \$20,000m. And to keep the oil business expanding, the oil industry as a whole is now spending some \$25,000m. a year, of which the majors contribute around one-third.

For the major producers to consider taking over the oil industry's assets – which amount to only \$8,000m. in the Middle East, or less than 5 per cent of the industry's worldwide investment – is one thing, particularly when their commanding position in the world's oil trade is considered. But for the consumer governments to take over from the oil companies and carry the burden of future investment in transportation, refining and marketing is quite another, even if they felt that they were capable of handling this end of the business. And as long as the major oil companies are left with most of the marketing assets, then the producers must deal with them at one level or another at the producing end.

The question is not only one of pure finance, but also one of management. Although some of the political advantages of having a commercially-operated oil business acting as a buffer between producer and consumer may have become rather less obvious over the last few years, there remain sound practical reasons for continuing to operate the oil trade on an international basis. This is not only because requirements of the major consumers must be balanced with those of the producers at an international level in view of the uneven speed of oil reserves around the world. It is also because the particular product requirements of the consumers have demanded an increasingly complex system of trade in crude oil and products to balance out particular qualities of crude oil with individual product demand and seasonal patterns of requirement as well as larger and larger units of transport and manufacture in order to ensure the lowest cost of oil trade in its various parts.

Main product yields from varied crudes
(using basic distillation, platforming and visbreaking processes)

	Petrol %	Middle distillate %	Fuel oil %	Refinery fuel and loss %
Bachaquero	5	19	70	6
Kuwait	13	38	42	7
Qatar	19	50	24	7
Light Seria	21	64	9	6

At one end, there is a wide variety of crude oil qualities. No two oilfields have exactly the same characteristics, still less do oils produced from different basins. Some, like the heavier Venezuelan and Kuwaiti crudes, tend to produce particularly high yields of the heavier oil products when broken down in the normal refining processes. Other oils, like the North African crudes and some of the Abu Dhabi oils, tend to produce much higher yields of the lighter products such as petrol and naphtha, while others are particularly suitable for 'middle of the barrel' products such as heating and diesel oils. At the same time, some crude oils are 'waxy' and require special treatment in distribution and refining, while others, such as the oils from Kuwait and the neutral zone, have particularly high sulphur contents, making them less valuable in pollution conscious markets.

At the other end, no two consuming countries have exactly the same pattern of product demand either. In the United States, for example, petrol sales account for some 44 per cent of the total 'barrel' against around 19 per cent for gas and diesel oils, 15 per cent for fuel oils and 7 per cent for kerosines; while in Western Europe as a whole petrol accounts for less than 25 per cent of total sales, while gas–diesel oils and fuel oils each account for around 33 per cent; and, in Japan, fuel oil, which is heavily used for electricity generation and industrial uses, accounts for about 50 per cent of the barrel against only 12 per cent for gas–diesel oils and 22 per cent for petrol. Even within Europe there are wide differences in product demand. The United Kingdom, where oil has a particularly strong position as an industrial fuel, consumes nearly 40 per cent of oil products in the form of fuel oil compared to 28 per cent for petrol and 20 per cent for gas–diesel oils; in Germany, where oil is in especially high demand as a heating oil, the middle distillates account for nearly 60 per cent of the barrel compared to less than 30 per cent for fuel oil.

Main product demand patterns, 1973

	United States %	United Kingdom %	Western Europe (excl. UK) %	Japan %	India %
Petrol	44	28	23	22	17
Kerosines	7	7	3	7	22
Gas oil–diesel fuel	19	20	33	12	28
Fuel oil	15	39	33	50	23
Others	15	6	8	9	10

And these national patterns are further complicated by the seasonal variations in the pattern of trade. Petrol consumption reaches its peak in summer, when holiday motoring comes to the fore, while heating oils, which have shown an especially strong growth in Europe over the last decade, reach their peak in winter

and are clearly highly sensitive to the impact of weather conditions on the rates of demand. The growth in consumption for
individual products may also vary widely from year to year and
from place to place depending on the specific circumstances of
industrial growth in each country, the failure of alternative
sources of power such as hydro-electricity after a dry spell or
government actions to curb pollution or change the fuel burn
of power stations at any one time.

These variations of crude oil qualities and demand patterns
are in many ways dealt with best on an international scale, so
that the imbalances of time and place can be made up from the
surpluses in other areas and the variations of product demand can
be met with a variety of crude oil sources. And it is this that the
major oil companies have increasingly provided over the last
decades, with the network of pipeline facilities, the creation of
large export refineries and the development of central import
facilities in Western Europe and elsewhere. If nothing else, the
crisis of 1973 showed just how complex the pattern of European
and international oil trading had become and, as individual
countries started to bring in export controls of products and the
effects of the Arab embargoes on Rotterdam began to be felt by
a wide variety of neighbouring countries dependent on the port
for their supplies, how difficult it is to tamper with it.

The same reasons also provide a strong practical justification
for the other main characteristic of the international major oil
companies – their vertical integration through all phases of the
business from production to petrol retailing. It may be that one
of the driving forces for this integration has been the simple
desire of the major oil companies to find an outlet for their vast
Middle East oil reserves and that this – as some critics have argued
– has given them an unfair advantage over their competitors.
But even if oil reserves had been broken up into much wider
ownership or had been taken over by the producing countries
much earlier on, the demands of the oil trade would have tended
to create large, multi-national integrated oil companies as the
most effective way of balancing demand against supply, financing
the growth, backing the trade with sufficiently large research
effort and, not least, obtaining a long-term security of supply and

sales. If the international oil companies have been most successful
at developing integration, it has nevertheless been the same forces
which have impelled national companies like ENI of Italy and
the Japanese and German groups to attempt to do exactly the
same.

The position of the majors will undoubtedly come under
increasing pressure over the coming years. With greater freedom
to gain new crude supplies directly from the producers, the competition from the national companies will become more intense
while the consumer government move to gain greater control of
the oil trade is bound to mean stronger support for these groups.
Nor is there any final reason why the major companies should be
the sole repository of expertise in any one area of activity. For
reasons largely of finance, the international companies have
increasingly tended to use the services of contractors for their
drilling operations, for the construction of their refineries as well
as agents for selling heating and fuel oils and the tanker market
for hiring rather than owning their ships.

Refining capacity in major areas by type of company
(excluding North America and the Communist world)

	Major companies		Independent companies		State companies	
	1960 %	1972 %	1960 %	1972 %	1960 %	1972 %
Western Europe	65·3	58·9	26·3	27·2	8·4	13·9
Africa	49·7	40·6	28·3	31·3	22·0	28·1
Middle East	76·4	57·0	9·8	11·6	13·8	31·4
Far East	28·9	29·7	47·9	59·0	2·4	11·3
Oceania	87·7	74·0	12·3	25·4	—	—
Caribbean	77·9	73·7	1·8	7·6	20·3	18·7
Other Western Hemisphere	24·0	18·8	10·3	8·5	65·7	72·7
Total	65·6	52·1	19·6	29·4	14·8	18·5

Government, national companies or independent companies
could hire the same services should they wish to enter these fields.
But the management of the trade as a whole, the raising of funds
necessary to finance it and the flexibility of supply and outlet
needed to balance it remain less easy to tackle. The more complicated the business becomes with pollution controls, with more

direct intervention by government in the organization of the market as a whole and with the need to explore in more and more difficult areas of the world, the more that governments are likely to call on existing skills to assist them rather than attempting to start from scratch themselves. This is already being seen in Algeria's enthusiasm for a return of the international oil companies to exploration there, in Iraq's negotiations with a group of European companies for exploration and agricultural assistance there and in the several negotiations going on in the Middle East over the construction of new refineries and petrochemical plant. And it may also be true for the consuming governments, whose primary concentration may be directed towards the development of coal and nuclear power and whose interest, therefore, in oil developments may be confined to supervision rather than actual management.

Under these conditions, the role of the international companies will become more complex and less dominant. In perhaps large part they will continue on as before, organizing oil trade through an integrated pattern of international sales – although their share of the trade will clearly fall at least to their level of refinery ownership and probably less than this as their position is gradually eroded by national companies over the future. In part they will become much more an agent of governments both in supply and in particular aspects of the trade such as oil exploration or refinery construction. And in part they will act much more as free market traders in oil as they seize particular advantages offered by the growth of an open market in crude oil following the state take-overs of production.

Almost certainly this will involve a greater decentralization of the major oil groups into national subsidiaries than before, a trend which will probably be further encouraged by the fiscal actions of governments to make them more readily subject to national taxes. Almost certainly, too, their position will be profoundly affected by the market conditions of supply and demand. If oil remains a very tight commodity with shortages ever present on the horizon, then governments will tend to take a much more direct control of operations to assure themselves of supplies. If supplies ease, on the other hand – as they look like doing at present – then the pressure for governments to intervene in the organization of the trade will be less. The point is simply that as long as the international oil companies continue to handle the bulk of oil sales in the market, they remain valuable to the producers, and as long as they are favoured by the producers, they remain valuable to the consumers. Particular crises of supply or demand could encourage either the consuming or producing countries to 'cut them out' of the trade at either end, in which case their value at the other end declines. But, unless and until this happens, they are probably too convenient and firmly established as the main part of the structure of the industry for either side to consider easily doing without them altogether.

But as the industry enters the new political and trading conditions of the 1970s and 1980s, the concern of the international companies may be less whether governments will allow the continuance of the vertically-integrated, international structure that has so much dominated the business during the past than whether the financial conditions of profit and return will be favourable enough for them to continue this role.

CHAPTER 26

Profit and competition

'Adequate energy is dependent upon adequate investment. That investment will only be made if there is some degree of confidence within the enterprises responsible for the investment on the shape and stability of energy policy.'

*Sir Eric Drake of BP**

The oil companies' constant pleas of inadequate returns and soaring costs have never struck a particularly responsive chord with the public at large. Ever since the days of Rockefeller and Deterding, the industry has had a reputation for vast wealth and ruthless materialism, the image of a sort of giant carnivore stalking the energy world, devouring its opposition, trampling over national controls set to confine it and pursuing its quarry with a ruthless desire for ever greater size and wealth. As long ago as 1887 a director of the Great Standard Oil Trust wrote to Rockefeller:

Our name is known all over the world and our public character is not one to be envied. We are quoted as representatives of all that is evil, hard-hearted, oppressive, cruel (we think unjustly) but men look askance at us, we are pointed out with contempt, and while some good men flatter us, it is only for our money.

Time may have softened this picture, and it has certainly changed the face of the industry radically from those days, but the image of Croesus, albeit in corporate rather than individual form, has remained with the industry ever since.

However unfair the oil companies may regard this, and however far from reality the popular conception of the industry may have

* Speaking at the *Financial Times*' Conference on World Energy Supplies, 18 September 1973.

become, it is hardly surprising. When one single oil company, Exxon, can make profits of nearly $2,500m. in a single year, can achieve a total sales revenue of $28,500m., the equivalent of the entire NATO defence budget, and spend something like $100m. solely on research and development of a new corporate name, 'Exxon', the man in the street can be forgiven for believing that this is power and money unequalled.

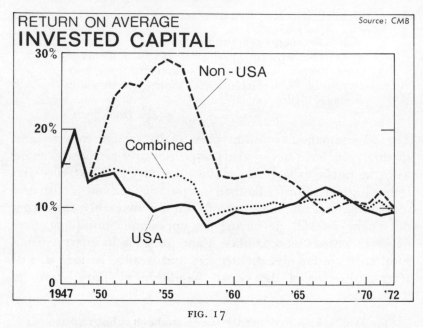

FIG. 17

Oil is big business. The seven major oil companies alone handled nearly 30m. barrels per day out of a Free World total of around 4·8m. barrels per day in 1973. In the same year they refined nearly 23m. barrels per day of oil, produced nearly 30,000m. cubic feet per day of gas, sold over 24m. barrels per day of oil products, recorded a net income of over $8,500m. and reported a total value of fixed assets of over $50,000m. On this scale of operations, an increase in profit of only a fraction of a penny per gallon can make the difference of tens of millions in a company's overall income, just as a fall in profit per gallon of only a fraction of a penny can produce an equivalent reduction in over-all income. And when figures of this kind rub off on individuals, like tanker owners at a time of high freight rates, or the wildcatter who strikes a large field in the United States or even the whore in a boom oil town in Alaska, the sudden wealth can be enormous. The Gettys, Hunts, Onassises and Niarchoses are all part of a business where money moves in huge sums and moves extremely quickly when new finds like the North Sea present new opportunities.

But money is only meaningful in terms of what it buys, and it is in the context of what the industry and the oil companies will have to invest over the future on new exploration, new distribution and new marketing facilities to ensure the necessary continuance and expansion of supplies over the future that the real worries begin to arise. Over the fifteen years from 1955–70, the industry is estimated to have invested over $100,000m. in the search for oil and gas in the Free World, and a further $115,000m. was spent on bringing it to market. Over the future the costs can only rise. Inflation is now surging forward at rapid rates of anything between 10 and 20 per cent per annum through most of the world. Pollution and environmental safeguards are adding substantial sums to the cost of developing almost every sector of the trade. The search for new oil and gas is taking the industry into ever more difficult and more expensive areas of the world. Demand, although faltering after the dramatic price increases of 1973, is still continuing on and is likely to grow still further over this decade at least.

Estimates of just how much money will be needed to fund future growth vary considerably, depending on what calculations are put into the sum for inflation rates, consumption increases and production costs. Figures for this decade suggest a total investment requirement of anything between $4,000–8,000m. and a recent calculation by the Chase Manhattan Bank has estimated that, if demand is to double over the period 1970–85, a total capital investment of over $800,000m. will be required – almost four times the amount spent in the preceding fifteen years. Of this over half would have to be spent on exploration and production while the industry would also require perhaps as much as $500,000m. again for the servicing and repayment of debt and the payment of dividends to its shareholders – making a staggering total requirement on the industry of around 1·4 trillion dollars.

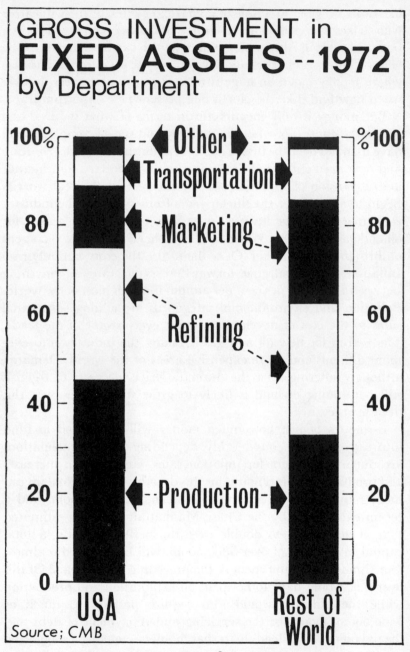

FIG. 18

This sum could well be exaggerated in view of the likely reduction in consumption growth rates previously predicted. But even assuming a much lower figure for the growth in demand for oil, and hence the tankers, refineries and distribution facilities needed to supply demand, the figures are still likely to be extremely high. The high price of Middle East oil and consumer government demands for a more rapid development of alternative crude oil sources from tar sands, shale, from secondary and tertiary recovery of oil in existing fields and for more intensive search off-shore and elsewhere will probably add greatly to the cost of exploration and production. This area alone may require an investment of $150,000–250,000m. over this decade and perhaps $400,000–500,000m. over the period 1970–85 as exploration and development costs outside the United States, which averaged around $30–35 per ton a year of new capacity in the 1960s, rise to double that figure or more. Estimates for the amount needed to be spent on refineries and other plant may have to be revised downwards in the light of changed demand figures, but previous estimates of inflation rates at around 4–5 per cent may well be far too low. Most oil companies are now building in figures for nearer 10 per cent inflation. Unless something very drastic occurs to change the picture of rising costs and higher interest rates that now presents itself in the West, this means that the investment in 'downstream' over this decade is likely to run to several hundred thousand million dollars just to keep the system ticking over on even a low rate of growth. Under all these conditions it might still be reasonable to assess the industry's total financial requirements at around $500,000–600,000m. during this decade (including debt servicing, dividends, etc.) and around $1,000,000m. for the period 1970–85.

It is against this background of financial need that the industry's profits and returns must be set. And the picture undoubtedly does give rise to concern. To achieve the kind of rates of capital investment required above, the industry as a whole would have to increase its net profits at the rate of something like 15 per cent or more per annum and raise its levels of return on net assets to a similar figure. But the trend of the last decade, aside from short-term spurts in profitability at times of particular crisis such as

1971. 1973 and 1974, has been for a steady decline in the rates of return to set in. Starting from an admittedly high base during the 1950s of an average return on invested capital of around 14–15 per cent, the rate of return fell almost consistently from the late 1950s onwards to a level of around 11 per cent and, despite some growth in the middle of the decade, went sharply down to less than 10 per cent during 1972.

The picture was particularly sharp in the Eastern Hemisphere, where the pressure of producing governments to raise their level of tax take by stabilizing tax reference prices at one end and the growth of competition feeding on a surplus commodity market at the other end both combined consistently to erode the profit margins of the large international oil companies. Despite an increase of over 12 per cent per annum in oil production, the seven major international oil companies managed to increase their net earnings by less than 7 per cent per annum through the 1960s and first years of the 1970s, while their earnings per barrel fell by nearly 5 per cent per annum from 54.3 cents per barrel in 1961 to 34.1 cents in 1971 and only 28.3 cents per barrel in 1972. Their return on net assets in the Eastern Hemisphere meanwhile fell from over 13 per cent at the beginning of the decade to less than 12 per cent at the end of the decade and only 10 per cent in 1972.

To a certain extent, the large oil companies, both the majors and the bigger international companies like Amoco, Continental, Atlantic Richfield and Phillips, were able to buttress low Eastern Hemisphere profits by buoyant United States' earnings, supported by the United States' system of import protection and production programming. Whereas in the 1950s the industry was gaining returns of anything between 20 and 30 per cent in the world outside the United States against returns of only 9–12 per cent within, by the mid-1960s these positions were being reversed. Returns in the United States climbed to 12 and 13 per cent, while returns in the rest of the world fell to 10 per cent and less. Companies with strong United States' positions, like Texaco, managed to balance one side off against the other, companies like BP or CFP with less strong positions in the United States suffered accordingly.

To a certain extent, as well, the large integrated oil companies were able to counteract the steady fall in product prices in the market by a sustained effort towards greater efficiency through economies of scale. Backed by a strong crude oil position in the Middle East and their financial muscle, the major oil companies in the 1960s invested on a grand scale to take full advantage of their integration in large refineries, combined pipeline distribution and even bigger tankers. Although this gave rise to the suspicion that the major oil companies were using their crude oil position and their international integration of operations to squeeze unfairly their competition, it was inevitable.

World refining, 1973

Crude oil processing capacity	million b/d	%
United States	14·2	22
Canada	1·8	3
Other Western Hemisphere	7·0	11
Europe	18·6	29
Africa	1·0	2
Middle East	2·6	4
Far East and Australasia	8·9	13
USSR, Eastern Europe and China	10·4	16
World total	64·5	100

During the 1950s and early 1960s, the majors had tended to treat their downstream operations mainly as an outlet for their crude oil production, where the profit centre of the business lay. Petrol stations, terminals and storage facilities and even refineries proliferated throughout Europe and elsewhere as the large international companies tried to get their brand represented in every important market, almost regardless of cost. And this concentration on spread rather than return left them peculiarly vulnerable to the impact of the newcomers who made their presence felt during the 1960s. Small independent retail companies began to cream off the most lucrative parts of the trade such as petrol retailing by concentrating their efforts on high-volume outlets, low-cost supplies and low outlets. Added to this competition, and fuelling it, were independent refiners who were able to take

advantage of cheap crude oil bought from North Africa and sometimes heavy state support for investment in developing regions
to establish processing capacity in countries like Italy; the state
oil companies, who could afford to operate on much lower rates
of returns than private enterprise, and, not least, the independent
companies like Continental and Occidental, who, having found
oil reserves of their own, were now anxious to integrate into the
downstream market themselves.

Faced with these kinds of competitive pressures – and far from
being uncompetitive between themselves – the majors' obvious
response was to use their own strengths of size and international
integration to achieve the maximum economies of scale. The
1960s became a period of constantly increasing investment on
the part of the international companies to achieve two aims;
one, to balance their crude oil supplies with their markets as far
as possible, and the other, to reduce costs by installing larger
units in almost every phase of their trade.

Companies like Shell and Exxon, with large marketing facilities
but 'short' on crude oil production to supply them, moved heavily
into new exploration in order to raise their crude oil resources.
Companies like Standard Oil of California and Continental Oil,
with large crude oil reserves but without large marketing outlets
to sell the production, did the opposite. At the same time everything was done to reduce unit costs by enlarging facilities and
pursuing the tightest possible integration of facilities. Mammoth
tankers of 200,000–250,000 tons deadweight became more and
more a normal part of the scene, particularly after the Suez Canal
was closed for the second time in 1967. And in order to take
advantage of this trend, major import terminals were built in
Rotterdam, Antwerp, the Thames and elsewhere and large
storage facilities and export refining plants were built around
them.

Refineries themselves were built increasingly as integrated
units, combining all the processes of distillation, vacuum reform-
ing, platforming and hydrocracking in a single complex, in order
to achieve the lowest unit costs and gain the maximum flexibility
of output to suit demand patterns. Companies combined together
to build large trunk pipelines to take both crude oil to inland

refineries and products to major distribution centres. Europe and the major consuming areas became more and more of a vast integrated market in which oil products were interchanged and facilities shared in order to balance out variations in demand patterns, ensure as large a scale of trade in oil wherever possible and in order to ensure the greatest flexibility possible on supply and product movements.

World's tanker fleet, end 1973

	Existing		On Order	
Size groups (dwt)	Number	'000 dwt	Number	'000 dwt
10,000–49,999	2,118	55,406	267	7,639
50,000–99,999	681	47,342	177	14,424
100,000–149,999	152	18,020	200	25,424
150,000–199,999	41	7,160	42	6,665
200,000–249,999	244	54,345	96	22,239
250,000–299,999	100	26,460	223	59,087
300,000–349,999	10	3,248	64	20,413
350,000–399,999	1	367	37	13,750
400,000 and over	2	952	65	27,990
Total	3,349	213,300	1,171	197,631

At the marketing end, too, the international companies were prodded by increasing price competition from the newcomers into paring down their own network of outlets and adopting similar philosophies of high-volume throughput marketing in the major growth areas. Although unwilling to go for price incentives on petrol in the same way as the cut-price operators, companies did start in the mid-1960s to follow much more aggressive point-of-sale techniques with the introduction of self-service and the adoption of trading stamps and promotions of collecting items, free gifts and games. The number of petrol outlets were reduced and retailing petrol, which had for so long lagged behind the developments occurring in other forms of retailing, at last began to follow the principles established by the supermarket in terms of lay-out design, offers and the consideration of 'own brand' tyres, batteries and accessories (TBA, as they are known in the trade).

The bulk industrial markets presented different problems. Here the key-word had always been service and spread of sales and this too was refined with the offer of 'packages' of servicing, back-up and equipment installation. But a lower value-added product than petrol, residual fuel oil, tended to be treated as something that had to be got rid of rather than earn profits in its own right and on this product, as on bulk sales of heating and gas oils to industry and commerce, it was price competition through high discounts of the formal price that dominated the scene through the 1960s despite the fact that the competition from newcomers was much less strong in these fields.

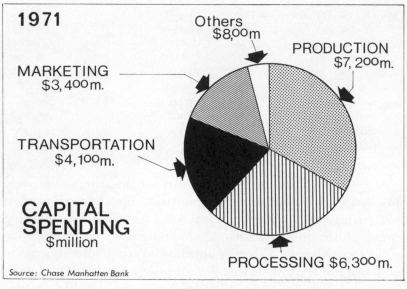

1971

Others $8,00m.

PRODUCTION $7, 200m.

MARKETING $3,400m.

TRANSPORTATION $4,100m.

CAPITAL SPENDING $million

Source: Chase Manhatten Bank

PROCESSING $6,300m.

FIG. 19

The move towards tighter integration of operations during the 1960s and ever-increasing economies of scale produced the picture – seen in other large volume industries – of increasing capital expenditure on the one side coupled by falling price on the other side as companies competed more and more to gain the full benefit of high-volume throughputs. On the political side it tended to raise fears in the market that the 'little man' would be squeezed out of the business altogether while among the producers it

15. A drilling site in Oman

16. Kuwait's oil port, Mina Al Ahmadi

aroused increasing friction between the desire of the oil companies to gain maximum flexibility of crude oil sources with the producers' desire to ensure that their own production was pushed as rapidly as possible. But for all that, it did prove remarkably effective in keeping down unit costs in the oil business and actually reducing them through most of the decade.

By the end of the 1960s, however, it was clear that the whole trade was beginning to get into serious problems. Returns in the United States, which had so long buttressed the fall in profits of the Eastern Hemisphere, now began to fall at almost the same rate as in Europe. Competition to gain outlets was reducing the price of oil faster than investment in larger units was reducing the cost of refining and marketing it. And, most worrying of all, the benefits of increasing economies of scale were beginning to taper off. There remained, and still do, further economies that can be made by the greater use of pipelines and large-scale rail movements of oil in the market and by integrating oil flows more. But the big jumps in cost savings which can be made by going for larger tankers, refinery units and other plants do seem largely in the past, at least in the present state of technology.

Tankers are being built of 500,000 tons and there is talk of 1m. ton ships, but the reduction in unit costs of construction and operation are fairly small. There are serious problems of the routes that such large ships can follow and the ports which can take them. Some doubts have been expressed about the design and the considerable problems involved in insuring such large units, whose loss could involve such vast sums of money. And the same is true of refineries and other plant. Although it may be technically possible to pursue still further larger units, the risks of accident become much greater, the environmental objections increase and the benefits in terms of the saving per unit cost become less obvious. There are also increasing political and environmental objections to the 'clustering' of refineries and the construction of massive pipeline systems that have been so strong a feature of the past decade.

At the same time as this stage has been reached, the inflation rates in the West, particularly on capital plant, have rocketed at a rate which was inconceivable in the last decade when inflation was counted in terms of a few per cent at most per annum. Materials such as steel, wages and finance have all got to a state where rates of inflation in double figures have become the norm rather than exception. The effect on the oil industry has been a virtual doubling, and in some cases tripling, of capital costs in almost every phase of the business: 200,000-ton VLCC tankers, which had cost about $14m. when first introduced in 1967–68, by 1974 had risen to some $40m. The daily operating cost of an off-shore rig, which had averaged around $15,000 per day in the late 1960s, has now increased to around $30,000–40,000, while a semi-submersible rig that had cost $8.5m. in 1968 now costs well over $30m.

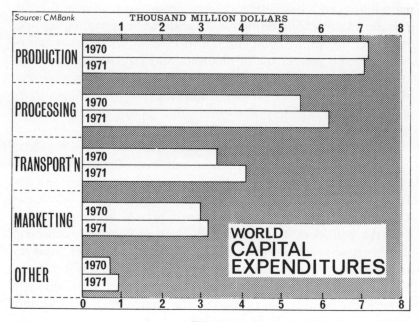

FIG. 20

The capital cost of new storage, at around $15 per ton in the late 1960s, by the early 1970s had risen to $25 per ton. The capital cost of new refinery facilities, at between $600 and 850 per daily barrel of intake in 1968, is now anywhere between $1,000 and $1,500, while the cost of items like road tankers, petrol pumps and pipelines has more than doubled over the same period. And this is quite aside from the additional costs that are arising because of the move towards off-shore exploration, where the costs of development in the North Sea for example now average some $3,000 per daily barrel of capacity against $100–300 on-shore in the Middle East, and the additional expenditure required for environmental reasons, government demands for greater storage and higher costs of insurance and servicing capital.

It is this picture of dramatically rising replacement costs plus greater expenditure on new exploration and development which probably worries the oil industry, like other capital-intensive industries, more than any other single problem. The sheer size of burden involved in keeping pace with expanding demands may, of course, be partly relieved by greater involvement of governments. There may be arguments as well for saying that the normal criteria of capital needs and rates of return tend to exaggerate the problems, although there is far more evidence that inflation makes a nonsense of profit and return figures assessed by traditional accountancy practices. But there are probably grounds for believing that some of the fears raised about the capital market's ability to meet the requirement for funds from the industry may be overstressed.

The rapid rise in prices for oil and the increased importance of energy to the Western economies will tend to divert resources from elsewhere into this field. Producer country surpluses will find their way back into energy investment in one way or another, although it is more likely to be indirectly through Arab investment in the capital markets generally rather than in the form of direct investment by the producers in oil-related enterprises in the West. The ability of the markets such as the Eurocurrency market to adapt to this kind of situation is well proven. The industry itself may well be able to allow a ratio of debts to assets much higher than it has traditionally felt willing to do, and its ability to raise large funds through 'off balance sheet' financing in the form of project loans is increasing, while the oil 'crisis' has done wonders for company returns, at least in 1973.

But heavier borrowing, whether on or off the balance sheet, does not resolve the situation. Indeed it actually intensifies the problems with time. Debts have to be serviced with interest payments at a time when interest rates are rising rapidly with inflation and the competition for capital from almost every industry. And, ultimately, of course, loans have to be repaid. Depreciation and the various methods of capital recovery which the industry has traditionally used to provide about a third of its funds are always vulnerable to political action by taxing governments, who could well act in a more restrictive way on this question over the future. On the profits side, the dramatic rise in profits and returns in the industry during 1973 still masks a five-year trend in which profits have barely kept up with inflation, never mind the increasing capital needs of the industry, and in which the ratio of debt to assets has increased to 30 per cent, and in the case of some companies much more.

To the extent that the consuming and producing governments want the oil trade to be carried on and financed largely by the international private oil companies – and there are strong reasons for believing that they do want this – then to that extent the oil industry will have to raise its levels of returns and earn enough profit to undertake the trade. And the industry's ability to do this will in turn depend on both the state of competition in the market and the degree of government intervention on prices and their finances.

Both these factors have been changed radically by the events of the early 1970s and both are highly unpredictable. In certain ways, the competitive factors that so dominated the 1960s have been reversed by the squeeze on supplies and the actions of the producers. No longer is it possible for independent companies to compete through the use of the cheap-owned oil that they once enjoyed in Libya and elsewhere. The Tehran and Tripoli price negotiations effectively ended the price differentials between the Gulf and short-haul oils and it is doubtful whether this price gap will ever be restored. The collapse of the surplus oil market in the early 1970s, too, has meant a severe weakening of the independent refiners and marketers who once used this surplus so effectively to compete on price incentives with the major oil companies. The dramatic developments of 1973 saw many of these independent small companies being squeezed hard on supplies. The great competitive market trends of the 1960s of forecourt promotions and ever increasing rebate-inducements on industrial oils almost disappeared for a time while competition on price gave way to the desire for security of supply through any oil company which could guarantee volumes.

But if this development has naturally aroused the widespread accusation that the major oil companies were using the situation to kill off their opposition and to ensure a cartel-type of domination of the market henceforth, the reality may well turn out to be very different from this. The international spread of the major companies and their vertical integration has always given them the appearance of much greater unity and co-ordination of competitive action than has in fact existed. Their similarity of structure has undoubtedly tended to make them think along similar lines about future planning and investment. Their drive towards maximum flexibility of supply and greater economies of scale has taken them into broad partnerships with each other, not only in the producing consortia of Iran, Iraq and other areas of the Middle East but also in the construction of joint pipelines in Europe and, in some cases, joint refineries and petrochemical schemes. It is also probably true that their position in the industry and the need to look at the longer-term future of the industry as a whole has made them more reluctant to indulge in price wars, for fear of their longer-term damage, than independent companies looking for rapid growth.

But in other ways there have always been wide and fundamental differences of attitude and interests between the companies. Some have had traditionally stronger and more profitable interests in the United States than others. Some have had more crude oil than they can refine and market through their own systems while others have always been short of crude oil. Some have had particularly strong marketing positions on specific products such as lubricants and aviation fuel oil while others have been stronger on petrol and gas oils. Some have large petrochemical or gas interests and others have not.

And these differences have found sharp expression in different marketing and commercial policies. What one company can withstand in terms of political pressure from individual producers or price competition in particular markets, another company cannot, and where one company may feel it in its long-term interest to establish a position in a certain market at the cost of short-term losses, another may take an entirely different view.

The variations in attitude and the competition between majors has been seen time and time again in the competition for bulk industrial contracts in specific countries, in the differences of approach towards the question of participation and the differences in response to the continuous demands of the Libyan government since 1970. If there had been no independent companies to prod the majors in most sectors of the business during the 1960s, the prices may well have remained more stable, but it is doubtful whether competition would have allowed them to rise in the way that Deterding, Lord Cadman and Walter Teagle wanted to see in the 1920s, or the way in which some of the majors had hoped to see in the immediate post-war years.

Now that the international oil companies are threatened with the loss of their traditional ownership of major crude oil reserves in the Middle East and elsewhere, these differences are likely to show themselves even more. In one sense it means that the whole industry must now try to move the profit centre of the trade to the consuming end and make its marketing and refining operations earn a much more effective return in their own right. And this must act as a strong discipline to any price competition.

But in another sense the removal of the traditional profit centre in Middle East and African crude oil production must act as a catalyst for greater competition in the marketing arena, particularly if a genuinely volatile crude-oil market appears. Companies that were previously 'short' on crude oil supplies for their marketing operations, and were thus forced to rely on long-term contracts at relatively high prices for a portion of their raw material, can now obtain those supplies much more easily on the open market and will use their strength in marketing to outbid their competition. Companies, on the other hand, that could previously afford relatively low returns in the market because of their pillow of crude oil profits will now be anxious to improve their standing in the market if they are not to be driven under by more efficient competitors, while companies that were previously satisfied to exist with a large measure of crude-oil sales will now have to enter the market in a more substantial way if they are to continue on in the business. And the picture may well be further complicated by oil discoveries, such as those in the North Sea, which will in turn encourage successful companies to enter the local market where they have not been in the past or expand their sales chains to ensure outlets for the oil, particularly if government policy insists that the oil be refined within the country.

Nor is it by any means a foregone conclusion that the position of independents, the cut-price retail marketers and the market brokers will be eroded in the changed circumstances. The crises of the early 1970s have been largely caused by marginal movements of oil supply in the market from relative surplus to relative tightness. A restoration of marginal surplus – which could well occur as demand slackens and output is restored in the Middle East – could easily reverse the situation. And this time the independents will have a substantial open market for oil to work on. Despite the fact that large volumes may continue to go through the system in 'buy-back' supply deals between the producers and oil companies and despite the fact that the major oil companies may retain some special access to crude oil under service or exploration contracts, the auctioning of short-term quantities of oil has almost certainly arrived to stay.

Because of the variety of conditions of demand and supply, too, there is always likely to be room for the broker and marketer operating on the surpluses in the market. And as long as there are these surpluses, and prices of both crude oil and products on the open market, however marginal, move up and down in response to specific circumstances of supply, tanker availability and demand patterns, then there will be price competition.

Even if the market circumstances remain tight and prices continue to rise, it is doubtful whether consumer governments will allow dramatic profits on the part of the oil companies. The same pressures of inflation that worry industry so much are also leading governments to take a more and more active role in controlling and supervising prices and profits of major commodities. During 1972 and 1973, virtually every Western government outside of Germany produced price-control regulations, and this intervention is likely to be further increased in the case of the oil companies by the growing role that consumer governments are now taking to influence trends within their energy markets. The industry's record profits in 1973 have already aroused wide criticism of their actions. At the very least, governments will probably act to reduce some of the tax incentives that oil companies have previously enjoyed and most will probably retain some kind of ceiling on returns or prices for a long time to come.

This will certainly provide problems for the companies, especially as the short-term political pressure for price restraint encouraged governments to stop price increases whatever the long-term effects on industry investment. But it need not be disastrous for oil companies. In so far as the oil groups do become either agents or partners of governments in production, transport and marketing of oil, they may well be allowed a reasonable return for the work, and indeed they will have to be if the investment necessary to sustain growth is to be forthcoming.

Past experience of government involvement in gas sales in Holland, marketing and refining in France, production programming in the United States and co-ordinated process plant investment in Japan all indicate that private enterprise can often operate in more stable and healthier financial conditions under government supervision than in a state of free competition. Shorn of their obligations to carry almost the full burden of oil industry on their own backs, oil companies may actually be better off with greater state intervention, especially if the returns of the national oil companies and the price of government-negotiated oil supplies are used as the yardstick for that supervision.

Yet, for all the changes that will undoubtedly occur with increased government involvement in the oil trade and the loss of traditional crude-oil ownership by the companies, one's suspicion is that marketing conditions for a sizable amount of the trade will continue on very much as before, with the rewards going to those who can most efficiently operate an integrated system of oil trading and those who can most successfully react to the cycles of freight rates and free-market oil prices. The one important new factor in the trade may simply be the degree of objectivity which companies will have to employ in considering investment opportunities now that they have lost much of the financial cushion previously provided by crude oil reserves.

Under conditions where, despite increased government involvement, companies will still have a wide choice of capital projects before them and where the cost of these projects will be enormous, financial return must become the guiding light of action. The 1960s showed all too sharply the penalties of a marketing philosophy based on the idea of growth for growth's sake. The 1970s may show that the days when the international oil companies felt obligated to maintain a market share of world trade whatever the cost are over. Already BP and Shell's withdrawal from the Italian market and Gulf's sale of its assets in Germany in 1972–73 have given some indication of the changing mood of company thinking. It is a mood that could do much to alter the face of the market over the future.

CHAPTER 27

Back to nature

'The problem is we can't scuttle the environment and
we certainly can't stifle economic growth.'
*Thornton Bradshaw, Atlantic Richfield**

Of all the pressures on the oil industry over the next decade, not
the least will be the pressure of popular concern about the
environment. The intense worries about the effect that the current
exploitation of fossil fuels and the high rate of energy consumption
is having on the ecology, or natural balance, of the world's
environment is one of the phenomena of our times. Combining
all the passionate hatred of the consumer-orientated society of
Western youth with the despair about change on the part of the
middle aged with the uncritically-accepted pronouncements of
concerned scientists, it has already forced politicians to revise
their ideas about what the public wants from economic growth,
brought about a profound change in the way that economists
have looked at future prospects for industrial development and
has had direct and sometimes extreme effect on the development
of the oil industry from the siting of its refineries to the rate at
which it has been able to proceed with exploration.

The roots of the current concern about the environment are
deep and complex. In part they are based on the very real, and
well-documented, worries about the impact of major oil slicks
on marine environment, about the effect of atmospheric pollution
by refinery, chemical and power plant on the atmosphere and
about the damage that can be wrought on surrounding animal
and plant life by excessive use of fertilizers, detergents or pesti-
cides. In part, too, they are based on less well-documented but

* In an interview with *Forbes* magazine, 1 January 1974.

nonetheless real concern about the potential impact on human life as a whole on the build-up of thermal and atmospheric pollution, and the depletion of natural resources, if present trends are continued. And in part the movement also reflects a much more intangible and generalized feeling of disaffection with the industrialized way of life.

For some, the environment has become a handle with which local communities can beat the head of centralized government and prevent the siting of refineries or power stations in their own communities, whether they accept that there is a national need for such plant or not. For others it has become a way of expressing a desire to return to different and older values of small, self-reliant communities – a 'back to the earth' movement. For still others the energy question is simply part of a much broader moral distaste for consumer society as a whole and the feeling that it must change if humans are not only to be safe but also happier – a feeling more broadly felt by those who have enjoyed the benefits of a consumer way of life than those who have not.

It is this amalgam of the moral, the scientific, the economic and the just plain outrageous which makes it so difficult to define precisely where the industry's own responsibilities lie on environmental matters and how far its future development will be influenced by such considerations. There are obviously areas such as marine and atmospheric pollution from oil, or the use of oil in combustion, where the industry can and certainly should carry a major burden of developing systems and technology for reducing both the causes and the effects of pollutants. There are equally areas such as the control of emissions from cars, the development of more efficient energy technology and the effort to reduce consumption, where the industry will not be so directly responsible on its own but where its activities and development will be very directly affected by the actions of governments and others.

But where the industry stands in the great moral and apocalyptic visions of those who believe that fossil-based energy systems of any kind are leading the world down the road to disaster is an unanswerable question. Issues such as the blanket condemnation of consumer society in which oil is an integral part are really matters of philosophy, not policy. Questions such as 'heat pollution' – the fear that the use of energy to create power will raise the earth's atmosphere to the point of suffocating man altogether – are not only totally unproven but would logically demand the doing away with power altogether. Other scares raised, such as the ultimate effect that carbon dioxide (which absorbs the sun's infra red rays) might have on heating up the earth's atmosphere are not only based on debatable evidence but are also founded on the extrapolation of past trends along a future line which is most unlikely to occur because of changes in energy technology that are already going on and others which are likely to arise well before the danger point, if it exists, is reached.

This is not to say that even the wilder visions of the 'limits of growth' followers are not valuable in pointing out where things might end up if nothing is done, nor that other of their arguments are not all too real. But the blanket approach of 'all industrial progress leads to disaster' that has so filled the public consciousness of late has had the unfortunate effect of diverting attention away from the specific realities of pollution facing the world today. And it is just these specifics which industry and government need to be charged with.

In the oil industry's case, the most obvious environmental problem is simply that of pollution by crude oil or oil products leaking from ships, storage vessels or production wells. The potential dangers involved in the increasing scale and volume of oil operations in the last twenty years have been amply illustrated by the dramatic sinking of the *Torrey Canyon* tanker off the south-west coast of Britain in 1967 and the leaks from off-shore oil production in the Santa Barbara Channel off the coast of California two years later, while the impact of persistent pollution from oil or oil combustion is all too obvious on the beaches of the Mediterranean, the inland waterways of Europe and the decaying fabric of innumerable monuments and buildings all over the world.

But it is something which oil companies, impelled by governments or not, can do – and have already done much to prevent through the installation of equipment throughout their operations to ensure immediate shut-down in case of leaks, the introduction of systems designed to reduce to the minimum any oil discharge with water and the imposition of procedures intended to ensure maximum safety in operations and the minimum possibilities of human error.

At the producing end, especially off-shore, the measures to prevent the possibility of a serious 'blow-out' or leak occurring, such as has happened off-shore the United States on several occasions, include a series of shut-off valves and blow-out preventers in and on top of the hole which automatically closes down the well should changes in pressure occur; the use of special drilling muds and well-casing to ensure that oil and gas do not break through from one formation to another; and the installation of additional safety equipment on the drilling platforms to flare off extra gas and to prevent the discharge of chemicals or oil into the sea. Similar 'shut-off' equipment is also used in underwater pipelines, where the line can be closed down immediately a drop in pressure is registered at either end, and the underwater lines are usually further safeguarded by burying beneath the seabed and by coating with a thick skin of concrete to give them extra weight and strength.

During the transportation stage by tankers, most international oil companies have now adopted the 'Load on Top' principle to prevent the greatest single source of marine pollution from oil, the washing of dirty cargo tanks at sea. Under this system, the oil washings from the cleaning operations, instead of being discharged straight into the sea as was customary in the past, are stored on board in slop tanks. The new oil cargo is loaded on top of this residual oil and discharged together at the port of destination. Further to this there has been much recent discussion about the possibilities of equipping new tankers with separate water ballast tanks to minimize the mixing of oil and water – although this would be a very expensive and long-term measure.

On-shore, the same kind of concentration on procedures and the installation of safety equipment is also being applied to the problems of discharge of dirty water from refineries, the potential damage that might occur from explosions in tank storage and terminal facilities and, not least, the discharge of oil into water during loading and unloading at tanker terminals. To back up their words with cash, the international oil companies and the tanker owners have recently set up two major insurance schemes to assure prompt and adequate compensation for those suffering from the effects of pollution in the Tanker Owners Voluntary Agreement Concerning Liability for Oil Pollution (TOVALOP) and the Contract Regarding an Interim Supplement to Tanker Liability for Oil Pollution (CRISTAL), which together will assure compensation up to some $30m. for any single pollution incident. In addition to this, there is a growing tendency to set up insurance schemes amongst operators in various off-shore ventures such as in the North Sea and a growing tendency also for operators in particular ports to band together with the local authorities and adopt procedures for the immediate clean-up of oil slicks and leakages.

No one could pretend that this kind of development is the result of pure goodwill on the part of the oil industry. It has taken a number of major accidents like the *Torrey Canyon* incident to wake up industry and governments alike to the inherent dangers of an oil trade that has grown at such phenomenal rates over the last decade and whose units of transport and manufacture have grown so large. Nor would it be wise to pretend that there can ever be any absolute guarantee that accidents will not happen either through human error, the unexpected situation or the failure of companies to ensure the highest standards.

A well blow-out could always occur because of the peculiar nature of the geology of a specific structure or the use of the wrong drilling muds during drilling. In the rough waters where exploration is now taking place, it could take a long time to get at the well and seal it off. In the same way, the growing scale of tanker shipments, which now account for the movement of over 1,500m. tons of oil a year, is bound to raise the risk of collisions, explosions or sinkings involving very major pollution risks in view of the size of tankers now plying the narrow waterways of the English Channel and other crowded sea-routes. To this has been added the increasing proliferation of producing structures off-shore which, however well lighted and however well marked on charts, raise the same navigational hazards.

And it is in the area of marine pollution that the technological problems of pollution prevention are still perhaps greatest. The last five years have seen increasing research and development of chemicals to break down oil at sea, of methods to sink the oil and equipment to 'corral' the oil and pick it up before it spreads. Some are very promising but so far no system has been proved absolutely effective for clearing up oil quickly and completely on rough seas. Although research into specific incidents in Louisiana and elsewhere has shown that accidental spillage may not have the permanent effect on the ecosystem that some have believed in the past, there is no doubt that the introduction of large and persistent volumes of oil into the sea, coupled with all the other untreated sewage and chemicals that get pushed into it in the blind belief that the sea will absorb anything, is presenting a more and more serious problem to the spawning grounds of fish and the ecology of marine life as well as birds. Even if the oil industry is not responsible for all of it – engine oil from small boats remains a particularly virulent offender – it must bear much of the burden for clearing it up.

Nevertheless, on the principle that prevention is better than cure, there is no technological reason why a surprisingly high proportion of the current causes of marine oil pollution should not be reduced, if not virtually stopped altogether, by the universal adoption of systems such as 'Load on Top' for keeping oil residues on board ships, by the general adoption of all the safeguards and the procedures that exist for off-shore oil production and by a general investment in the safeguards that exist for keeping refinery operations cleaner, quieter and less smelly. What is needed is both greater research into the effects and ways of clearing up marine pollution and, even more important, an international determination to enforce the standards that can be achieved.

It is here that the greatest failings have been shown up. Despite successive meetings of the Inter-Governmental Maritime Consultative Organization (IMCO) and the adoption of an International Convention as early as 1954, laying down various safeguards intended to reduce pollution at sea, implementation of the regulations by national governments has been slow and far from universal, just as attempts to enforce reasonable navigational equipment and standards of seamanship have been less than completely fruitful. Until such international efforts do produce real international agreement and enforcement, the fruits of present pollution technology will always be less than they ought.

Direct pollution by leaks of oil is an issue which has long captured the public awareness and, as such, has seen a fair degree of general agreement between the more responsible elements of the oil industry and governments over what should and can be done. It is, after all, in the direct commercial interests of a company to see that accidents do not occur to its oil as it is being produced and transported across the high seas, and most companies must now recognize the damage that can be done if public opinion is aroused by even relatively small cases of oil pollution.

Less obvious to the public but more worrying to officials, and arousing perhaps greater potential conflict between governments and the oil industry, is the issue of atmospheric pollution, caused by the burning of hydrocarbons in refineries and in their end-use at the factory, the office and on the roads. The industry has already done much to improve the effluent problems of refineries through the use of air cooling, the recirculation of cooling water and various methods of gravity and chemical treatment to remove oil before water is discharged into waterways or the sea. Noise has been reduced through the installation of improved equipment, while the emission of odours, although still far from fully solved, has been reduced by minimizing the evaporation of oil and gases and restricting their release. But air emissions – basically sulphur dioxide, nitrogen oxides and unburnt hydrocarbons – has aroused greater argument over whether it should be handled at source, for example by reducing the sulphur contents of fuel oil products before they are used, or whether they should be controlled at the point of end-use, for example by the installation of high stacks at factories or anti-pollution devices on cars.

At the moment the main method for preventing the excessive ground concentration of sulphur dioxide from either refineries, power plants or factories has been by the use of high chimney stacks, which have proved reasonably effective in dispersing the emissions away from the immediate vicinity of the plant. Nitrogen oxides from the combustion in factories or refineries – one of the principal factors in causing smog under certain atmospheric conditions, as in the United States and Japan – can be reduced by refinements in the combustion systems. But there are increasing signs that the use of high stacks merely solves one population's problem by dispersing it to another area, and there is growing evidence of the harm that long-distance movement of sulphur dioxide in particular can have through acidification of the rainfall.

The problem is really one of cost. The sulphur content of fuels can be reduced substantially at source by the introduction of catalytic hydrogenation plant in refineries, and there are methods of desulphurizing flue gases as they go up the chimney. But the investment in the plant needed would be high. One estimate has suggested that to desulphurize residual fuel oil in the United Kingdom from an average 2·6 to less than 1 per cent sulphur content would cost as much as £200m. in capital expenditure and a further £100m. a year in operating costs, while the problems and costs involved in desulphurizing flue gases would be even greater.

The question is basically whether the money is best spent this way or another. For some in the industry, the case for proving that it really is necessary has yet to emerge and they point to the potential strain that sulphur controls on fuels might (and to some extent already have) bring to the supply and price of high-quality, low-sulphur crude oils. But others would argue that the risk involved in failing to move in this direction is too great for society not to invest in such measures. The United States, Japan, the Netherlands and Sweden have all adopted regulations setting out minimum sulphur contents of fuels to be reached by certain dates, and the publication by the World Health Organization recommending ultimate targets for levels of sulphur oxides is likely to lead to other countries adopting similar standards.

The choice between high pollution standards and lower cost and more easily-available fuel becomes even more apparent when pollution from motor vehicles is considered. The problems of photochemical smog from car exhausts, seen at their most extreme in Los Angeles in the United States and in Japan, where the meteorological conditions induce its build-up, have made almost every developed country extremely concerned about the problem.

But the issue is not an easy one even if it has become a popular platform. In the current state of engineering, perfect combustion in the internal combustion engine is never achieved and as long as it is not achieved, then quantities of carbon monoxide, oxides of nitrogen and unburnt hydrocarbons inevitably result and are discharged through the exhaust. Close co-operation between the oil industry and the car manufacturing industries has done much, at least, to solve some of the problems of evaporation of fuels into the atmosphere from the fuel tanks and carburettor. Improvements to the combustion system and better maintenance have also done something to reduce the emissions of oxides of nitrogen and other contaminants, particularly in the diesel engine.

The gasoline engine, however, has a long way to go before it can be considered totally unharmful. The question now posed by the rigorous legislation being introduced in the United States and being considered in Europe and elsewhere is how far and how fast can society move without creating impossible demands on the industry and enforcing practices which may create new pressures of their own in terms of additional petrol consumption. The classic example is United States' regulations, which now demand a rate of reduction in exhaust pollution so rapid that it can only be achieved by the installation of various devices such as catalytic reformers to ensure the burning of combustible materials in the exhaust. The efficiency and life of these catalysts, however, could be greatly reduced by the presence in petrol of lead additives, introduced to lessen the tendency of high-compression engines to 'knock' (produce uncontrolled combustion at high temperatures and pressures) and to enable engine compression ratios, and thus efficiency, to rise.

The amount of lead in petrol can be eliminated, but it can only be reduced at a high cost, with the addition of new plant at refineries and, most important, at the expense of requiring as much as 15–20 per cent more petrol to achieve the same performance. And the quandary over whether it is worth this cost is only made worse by the debate on the medical dangers involved in the build-up of lead in humans as a result of car pollution – a point which is certainly worrying health officials in major cities but has not yet been proved to be sufficiently intense to be actually dangerous to health.

The issue is not so much a question of absolutes – whether car pollution needs to be reduced or not – but one of degree. Ultimately the problem may be solved by producing better or different engines, either using low-pollution fuels such as liquefied petroleum or natural gases or running on completely different power sources such as batteries or fuel cells. Even in the more immediate future, pollutants from current types of engines could be substantially lowered by refinements in the internal combustion engine or a move to different power sources such as stratified-charge or other types of engines, which can provide similar propulsion power as the internal combustion engine without the same pollution levels. On the lead issue, too, time could enable the development of engines with lower requirements for lead. Or lead filters could be perfected. In the United States, the tendency has been to go for a rapid clean-up regardless of the problems, and it remains to be seen whether current worries about the availability of petrol will ease this pressure for the elimination of lead. In Europe, on the other hand, while most countries have now adopted targets laying down a reduction of lead content of 0·4 g/litre within a few years, there seems a greater readiness to accept the need for more modest immediate targets while pursuing the engineering solutions over the medium-term.

The question of automobile pollution and its causes also brings in the much broader questions of conservation of fuels and the efficiency with which they are used. Although the oil industry has tended to concentrate, and rightly so, on the specifics of oil pollution, the environmental movement has also made the public much more aware of the wider issues of the depletion of natural resources, the harm that uncontrolled development of energy can do and the need to look on industrial development not just in terms of its strict financial cost but also its costs in terms of environmental damage and depletion of natural resources. However wild this approach may have become in the hands of some of its proponents, it will undoubtedly have a profound impact on development – most especially in the area of conservation, where price and an all too real concern about the future availability of hydrocarbons have dramatically strengthened the case for reductions in fuel use.

Theoretically the scope for such savings is immense. Half the total volume of fossil fuels used in the industrialized world is wasted, mainly through the inefficient conversion of fuels into heat and power but also through the transportation of energy. Of the coal and oil used in power stations alone – one of the biggest single uses for fuel – less than 30 per cent reaches the final consumer in the form of useful energy because of the basic inefficiencies of electricity generation and the losses in transmission. Motor transport, another major user of energy, accounting for 15–20 per cent of total primary energy use in the West, wastes as much as 80 per cent and more of its energy input. Space heating by oil and gas has probably an average efficiency of 70 per cent in its energy use and for many industrial processes the efficiency may be no more.

Years of cheap fuel and plentiful supply, coupled with poor government planning and artificial price restraint on certain state-run fuels, have all too often allowed electricity, for all its wastefulness in the conversion of primary fuels, to compete with gas, which is a relatively efficient as well as clean fuel, for the heating market. Public transport has all too often been left to decline under the competition from the private car, despite the fact that congestion and urban driving increases the obvious waste of energy that comes with private transport. Consumers of all sorts have been encouraged, or certainly not discouraged, from putting emphasis on insulation of buildings to retain heat or greater efficiency in their processes.

Some idea of the potential saving that could be achieved by the introduction of systems such as total power, using the waste heat from power generation to heat buildings, from higher standards of insulation, more efficient combustion, the move to smaller cars and greater restraint on the part of the consumer can be seen from the estimates made by the Office of Emergency Preparedness in the United States in 1972. Adding everything together, the Office suggested that demand might be reduced by some 20 per cent in the United States below previous predictions by 1980 and as much as 25 per cent from its expected levels in the mid and late 1980s. This assumes, of course, a perfect response in every field, but even supposing that only a third of these savings was actually achieved, the reduction in consumption could still be as high as 1–1·5m. barrels per day of oil by the end of the decade and over 2m. barrels per day by later in the 1980s, and the predictions do appear to be borne out by the immediate impact on demand of the crisis of prices and supply of the winter of 1973–74.

Estimates of potential energy conservation savings

	Million barrels of oil per day equivalent		
	1980	*1990*	*2000*
Transportation			
Increased efficiency of engines	—	6·0	8·8
Increased use of public transport	1·0	1·4	2·5
Accelerated trend towards large jets	0·3	1·0	1·5
Residential–Commercial			
Improved insulation, construction etc.	2·7	5·3	9·1
Industrial			
Waste heat utilization	2·2	4·0	13·9
Re-cycle of steel and aluminium	0·8	1·2	1·8
Re-use of glass, incineration of paper	0·3	0·7	0·9
Power Generation			
Decreased average heat rate	—	4·5	15·1
Total	7·3	24·1	53·6
Percentage of total estimated energy demand	(4)	(9)	(13)

Source: *Estimates prepared by Exxon Corporation*

The United States is an extreme example, however. Japan, in contrast, uses fuels relatively efficiently, while countries in Europe have neither the luxury use of energy nor the *per capita* consumption that the United States has. Price itself will do much to ensure a more careful use of energy and the investment in more efficient ways of converting it to power and heat in factories, homes and offices. Governments themselves may be able to do something in addition by encouraging the adoption of standards of insulation in new buildings and the decentralization of electricity generation with total power schemes and, perhaps most significantly, the most rational distribution of fuels through the market according to their best use. The idea of 'energy budgeting', the calculation of costs in terms of energy use, could also do something to induce more rational transport policies for urban conurbations and long-distance transport and the substitution of materials, such as iron and plastics, with relatively low energy content for those with high energy content, such as magnesium and aluminium, which require 4–7 tons of oil equivalent to manufacture each ton of the material.

But the results are unlikely to be anything like as dramatic as the proponents of 'energy budgeting' sometimes suggest, unless there is a really major breakthrough in the field of technology. Years of intensive research into potential improvements in the generation of electricity from fossil fuels has produced surprisingly little in the way of results and, until fossil fuels can be substituted by nuclear power, the waste is likely to continue. Equally, the idea of switching back from private to public transport, while attractive and certainly useful in particular urban situations, has severe limitations in terms of its applications to large countries of relative low density of population like the United States and might well involve a degree of government control that is unacceptable in a democratic society.

Many conservationist developments such as total power, or the generation of heat and electricity from waste, will take time and will necessarily be limited in their impact because of the changes in industrial structures that they imply. Nor are the aims of conservation always compatible with environmental concerns. The substitution of coal for oil in power stations as part of a re-deployment of fuel use, for example, would not only create particular environmental problems in the mining of large amounts of additional coal but would itself involve the use of additional energy in reducing the pollution problem that would arise. The idea of using the waste heat of power stations to heat districts or fuel energy-intensive industries runs counter to the environmental desire to put large plants as far away as possible from populated areas and would involve additional use of energy for transport of the goods produced. The substitution of materials such as glass for plastics because of the waste disposal problem would again necessitate increased energy consumption.

In the field of the environment, just as in other economic fields, there remain no easy choices, despite the consumer's inevitable desire to have it all ways at once. Certainly much can be done to clear up some of the most obvious incidents of pollution arising from oil and energy consumption. The impact of the Clean Air Act in the United Kingdom following the London smog of 1952 is an obvious example of just how much can be achieved by controls and there is no reason why similar results cannot be obtained from controls on oil spills and atmospheric pollution from hydrocarbon systems. But the further society goes beyond specific aims, the more it will confront choices between cost and cleanliness, between energy conservation and environmental control and, ultimately, between the industrial way of life and a return to decentralized, self-sufficient agricultural communities.

It may be, as some in the oil industry have hoped, that the scares over energy supplies raised during the past few years will result in a swing of the pendulum of environmental concern back to a less demanding position. But it is unlikely that public opinion will ever return to the old assumption that economic growth and industrial development were *per se* the right goals of any developing society. Economic progress over the future will now have to take account of pollution and conservation, however much these may produce contradictions and problems of supply for the energy industries.

CHAPTER 28

The continuing search

'Oil companies *cannot* alone command the flow of oil
from producing areas. . . . But companies *do* command
expertise in the search for oil and production. It is they
who discovered the oil in the first place and it is they
who are still discovering new sources, beneath the sea
and in Arctic regions.'

G. A. *Wagner of Royal Dutch–Shell**

If one of the most important single developments in the oil
industry during the 1970s has been the loss by the major oil
companies of their traditional ownership of oil reserves, this does
not mean that the exploration and production of oil has become
any the less important a part of the industry's life. Indeed, the
reverse may well be true. The rise in prices, the political problems
arising from the Middle East's dominance of the world's crude
oil supplies and the consumer governments' determination to
develop fuel sources of their own have all made the search for
oil more widespread and ever more urgently required.

The contractual basis on which oil companies take part in this
search may well change. State companies may become increasingly
involved. But the fact that the international oil groups will take
part in the search seems almost indisputable. Even in the Middle
East, where the sheer size of the fields makes production a relatively simple and inexpensive operation, many governments, such
as those in Saudi Arabia, Iran and Abu Dhabi, may prefer to
keep the companies on, if only on a contractual basis, in view of
the stability of market outlets that the companies still provide
and the management function which many of these countries are
not yet capable of handling in full.

* In a speech to the Netherlands Chamber of Commerce, 8 November 1973.

Even those traditional producers like Iraq, Algeria and Libya, who have all taken a much more radical course in the seizure of full ownership of their oil concessions, have shown interest in getting the international oil companies back into exploration, although on different terms, and to use their experience and management ability on other projects such as liquefaction and transportation of natural gas, the construction of refinery and petrochemical plant and agricultural programmes. Other countries which have yet to find oil but could well do with it, like many of the developing countries of Africa, Asia and Latin America, are now proving even more anxious to get oil and gas exploration underway in their lands as they face huge import bills resulting from rising fuel prices and see the opportunities of gaining revenue themselves if oil is found.

The developments of the last few years have clearly ended for ever the days when this exploration and development work was done under the traditional concession system, where the oil companies obtained rights to own the oil found on a concession area in exchange for a royalty payment of 12·5 per cent on output and various 'sweeteners' in the form of cash bonuses on the signing of the agreement and the start of production and the offer of technical assistance in other directions. The end of this system may be no bad thing. While perhaps inevitable in the days when empires still existed and countries were glad to get oil developed at no cost to themselves, the concession system has become increasingly incompatible with a world where economic nationalism and accusations of economic 'neo-colonialism' by the West have become the order of the day.

The passing of this era has not been an easy one. As long as the old concessions of the Middle East held such a vast surplus of oil, companies were clearly reluctant to settle different and tougher terms on new concessions. British Petroleum and Exxon in particular long refused to accept any form of state participation, although such an acceptance would have given BP at least a concession in Saudi Arabia, for fear of the effect that it would have on their existing arrangements in the Gulf. But by the early 1960s this kind of approach was rapidly crumbling before the ambitions of the national oil companies, who were quite prepared to offer such terms, and used them as the bait to attract the producers; the independent companies, who looked at the prospects on more purely commercial terms and saw that state participation in Iran and other countries did not necessarily mean a lower rate of return on finds; and the international majors such as Shell who, short on crude oil, were anxious to develop new reserves.

By the end of the 1960s, state participation in one form or another had become the rule rather than the exception in the award of new concessions. By now it seems almost certain to be extended to most existing concessions as well. It has the obvious political attraction of making sure that the host government not only does get a fair share of the operations but is *seen* to do so. It establishes beyond doubt the state's right to share in decisions and it enables governments to take part in the fundamental question of how fast and how far their reserves should be developed – however much this may militate against the natural urge of the oil companies to suit output to the conditions of the international market.

The forms of this partnership can vary. Indonesia, after the fall of President Sukarno, has adopted a system of production-sharing, under which the state and the oil companies share oil production after a certain volume has been produced to pay off the initial capital expenditures involved. And this system has proved highly successful in attracting oil company investment in its off-shore development. Other countries, like Norway, have developed arrangements of 'carried state interest', under which the state takes no part in exploration but has the right to buy into a proportion of the licence once oil or gas has been found in commercial quantities. Other countries, such as Iraq in some of its more recent negotiations with oil companies, have preferred to set the oil companies on a purely contractual basis, financing the whole operation and paying the company a fee for their management. Libya has recently negotiated a deal under which the company (Occidental) carries out the exploration and development in return for a tax-free 19 per cent of oil found if output reaches a certain level. Another suggestion, much discussed in the United Kingdom, is for the government to buy all the oil, thus ensuring whatever rate of 'take' it wants in financial terms and also ensuring control over the rate of development and the destination of the oil.

Despite all the cries about 'principle' and 'precedent' which so dominated discussions in the past, the problem is not so much one of the particular form of the concession as the security of tenure and the precise financial terms attached to it. On the one side, governments obviously want concession agreements that are politically attractive to the public, which ensure them the maximum possible share in the proceeds and provide them with the means to control development at the rate best suited to their own internal needs.

On the other side, the concessionaires' main considerations must be reasonable security of tenure and a potential return, consistent with the risk and the investment involved. Whether the terms are attractive or not obviously varies from place to place depending on the geological chances of finding oil, the expense of producing it and the political and financial stability of the country itself. What may well be worthwhile on the traditional concessions in the Gulf, where the companies have largely repaid their initial investments many times over, where the geological conditions are unrivalled for their oil potential and where production costs can be as low as 6 or 7 cents per barrel, is clearly not worth the investment in a country where the potential oil basin is in the middle of the jungle a thousand miles from the coast, where the local government is liable to be overthrown any minute and where the geological surveys suggest only a remote possibility of finding oil.

The trouble over the past has been that, once oil has been found, everybody has tended to want the best terms going elsewhere in the world, whatever the particular conditions of their own markets, and this is as true of the North Sea countries and Australia, peering over their shoulders to see what is happening in the Middle East, as it is in the developing world. The concept of 'changing circumstances', so meticulously argued by Middle East lawyers as the legal justification for changing concession terms when developments make them appear less attractive, has tended to become too easily simply an excuse for extracting more and more concessions from the companies as soon as the actual discovery of oil and the expenditure on production facilities makes it difficult for them to withdraw from their investment.

This question of ensuring security of investment and a return on oil output commensurate with the risk prevailing when exploration was first started is unlikely ever to be fully solved unless new concepts of concession arrangements can be developed or the oil companies become entirely contractors. State participation will at least ease many of the political problems arising from foreign ownership of natural resources. An end to the concept of 'ownership' and its replacement by forms of production-sharing agreements will also do much to remove some of the difficulties encountered with the more traditional concession forms.

Average capital costs for oil

in £ per daily barrel of capacity

Middle East on-shore	150
Middle East off-shore	450
North Sea	1,200
Alaska	1,500
Tar sands	3,500
Oil from shale	2,500
Oil from coal	3,000

But the hard truth is that when no oil has yet been found in an area, the host government is prepared to offer considerable inducements for companies to come and try to find it. Once oil has been found, then the uncertainty is removed and the need to allow a return for risk becomes more difficult to defend in public. And this fact in turn encourages a situation in which companies seek terms which will allow them a very rapid recoupment of investment and thus create conditions where it becomes only that much easier for critics to claim that the country's birthright has been sold for a song.

In tackling this problem, the industry will also have to find a solution for the problem of who pays for the 'dry holes'. One of the *raisons d'être* of the international oil industry has always been that successful exploration in one part enables companies to support unsuccessful exploration in another part and hence sustain a widespread search all over the world. But this has become increasingly difficult to defend to the public of the country where successful oil production has started and who therefore see no reason why they should forego some of the potential 'economic rent' available to their nation in order to subsidize failures elsewhere.

A solution agreeable to all is not impossible, particularly if governments of newer producing areas are willing, as they may be, to allow a reasonably high rate of return on successful exploration once they are involved as partners in it. But it is often an underrated problem. Despite all the advances that have been made in the techniques for finding oil, exploration remains a gamble on a grand scale, where the stakes can run into tens of millions of dollars and where the chances of failure always remain higher than the chances of success. For every successful exploration well, a dozen prove failures. There are cases, such as the North Sea, where major oil finds are made where few would have dreamt of when exploration first started. There are other areas, such as off-shore the Gulf in the Middle East and in the Canadian Arctic areas, where exploration started off with the highest hopes and results have been so far disappointing. Shell has three times paid £20–30m. in bonuses or bids for off-shore acreage around the world and each time drilling has so far proved a failure, and the same story can be repeated in the offices of almost every international oil company.

Occidental made a fortune out of acreage in Libya that had been returned as worthless by Mobil. The successful NAM partnership of Esso and Shell in the Netherlands was founded when Shell, growing increasingly disillusioned with the results of its drilling in Holland, exchanged a half-interest in this venture with a half-interest in Esso's Cuban concession. Cuba turned out to be a failure, while further Dutch drilling found the Groningen gas field, one of the largest in the world.

British Petroleum, which is renowned for its exploration expertise and success, was still able to spend about £30m. on a twenty-year search in Papua without coming up with a commercially-exploitable find. The Shell–BP partnership in Nigeria, on

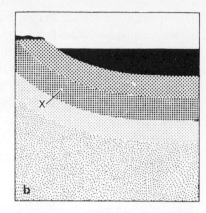

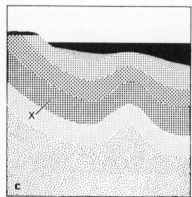

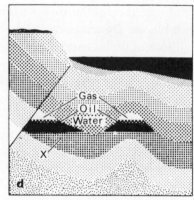

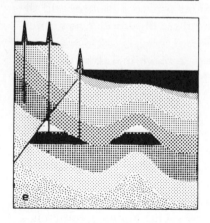

(a) *The remains of marine plants and animals settled to the sea bottom (x), mingled with sand and mud.*

(b) *Rivers emptying into the sea brought more sand, silt, and clay, covering earlier watery deposits.*

(c) *Bacteria or other agents turned the organic matter into oil and gas. Earth forces buckled the strata.*

(d) *Oil and gas migrated until stopped by impervious rock. They formed separate layers above the water.*

(e) *Wells are drilled to seek the buried formation. They may find oil or gas — or only water or a dry area.*

FIG. 21　The Formation of Oil

the other hand, spent an equal amount of time and an even greater amount of money but was able, almost at the point of abandoning the effort, to discover oil finally in what has since turned out to be one of the world's major oil-exporting countries. Even when oil is found, there is no absolute guarantee that production will fulfil all expectation. Techniques for evaluating the potential size have improved enormously so that a fairly accurate assessment can usually be made after a few wells. But hopes can still be disappointed, as they were with Denmark's first off-shore oil field, the Dan, where the reservoir conditions deteriorated radically soon after production started.

The uncertainties of exploration and development are inherent in the nature of oil formation and accumulation. Even the precise origin of oil is still not fully established. Broadly, however, it is now generally accepted that oil is generated from marine organisms which lived in their millions in the shallow waters surrounding the continents of prehistoric times. Under certain conditions of probably stagnant water, large volumes of dead organic matter drifted down to the sediments deposited by rivers on the seabed, where the fatty acids of which the matter was composed were transformed to petroleum compounds through a process of bacterial reduction.

With time, large masses of sediments containing this 'proto-petroleum' were built up. Then – again through a process still imperfectly understood but probably related to continuous heating over a period of time – this protopetroleum was 'cracked' into oil and gas as we know it. At this point, the oil could either stick where it was in small droplets held by surface tension, or it could, under the right conditions, be forced out of the source rock by compression, caused by further sedimentary layers accumulating above it, to an adjoining 'reservoir' rock, where the permeability and porosity of the rock was sufficient to allow the oil to collect in the interconnected spaces – rather like water in a sponge. After gathering volume there, a process of secondary migration could then take the oil on to other reservoir rocks where the oil could either escape to the surface in the oil seepages known to the world since ancient times, or it could become dissipated over a wide area or – as the geologists hope – it could become trapped

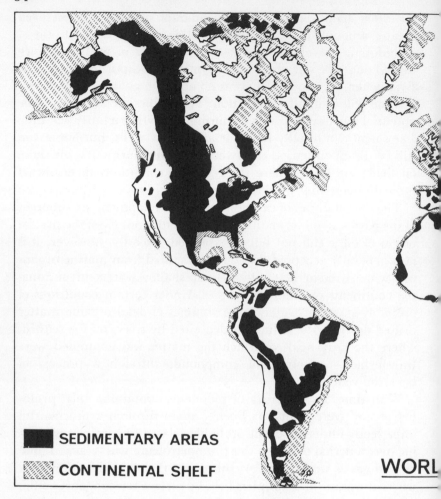

FIG. 22

and accumulate over time against 'caprocks' of impenetrable strata of salt, shale and clay distributed in such a way as to confine the oil in a large reservoir.

It is these traps, which can be formed in a number of different ways from simple dome-like folds to complex stratigraphic faults caused by changes in the characteristics of the rock, that the oil companies search for in exploration and whose presence is often difficult to predict. Sedimentary basins of porous and permeable

DIMENTARY BASINS

rock capable of allowing oil to migrate and collect are widespread, covering more than half the world's land area, and oil in minute quantities is found throughout them. But the right combination of source rock, adjoining reservoir rocks, final reservoir conditions and, above all, the right structures for trapping the oil, with the right time sequence in which all this could occur, is much less common.

Part of the problem is that the industry is still working backwards from what has been discovered to explain how it got there. The knowledge of oil generation and accumulation is still too incomplete to be able to predict accurately from initial geological or geophysical evidence whether oil will be around in commercial quantities. Increasingly sophisticated techniques of surveying potential oil-bearing sedimentary basins have improved the industry's ability to tell whether an area is likely to contain oil or gas. Geological field parties can tell from rock samples and fossils taken from outcropping whether the age and sequence of rocks is right for hydrocarbons to have been generated and collected. Geophysical surveys, using reflected sound waves from specially prepared explosions, can provide a great deal of information about the sequences and structural situations beneath the surface of the earth, especially now that computers have been brought in to interpret the sound-wave patterns.

But, although there remain hopes that further improvements in geophysical methods may enable geologists to define hydrocarbons in place from surveys, it is still only drilling that can tell whether oil has been actually generated in the area and whether it has in fact accumulated in sufficient quantities to make it worthwhile producing. And the chance element is even greater when it is remembered that the really large or 'giant' fields are few and far between but make all the difference on volumes and profitability. Some three-quarters of world reserves are accounted for by only 200-odd fields, in fact.

The uneven spread of these giant fields in a few really prolific basins of the world – at present mainly centred around the Gulf area of the Middle East, the Lake Maracaibo area of Venezuela, the Gulf coast districts and certain small basins of California in the United States, the North Slope of Alaska, Libya and, not least, the North Sea – has yet to be explained, and this makes predictions about future oil exploration extremely difficult. As the world's potential basins are more closely surveyed, the range of possibilities is narrowing. But a few major discoveries of giant fields could still alter the picture radically for the better just as failure to find such fields could depress it.

What the rapid rise in prices of the 1970s and the worries about security of supply of the last few years will do is to increase greatly

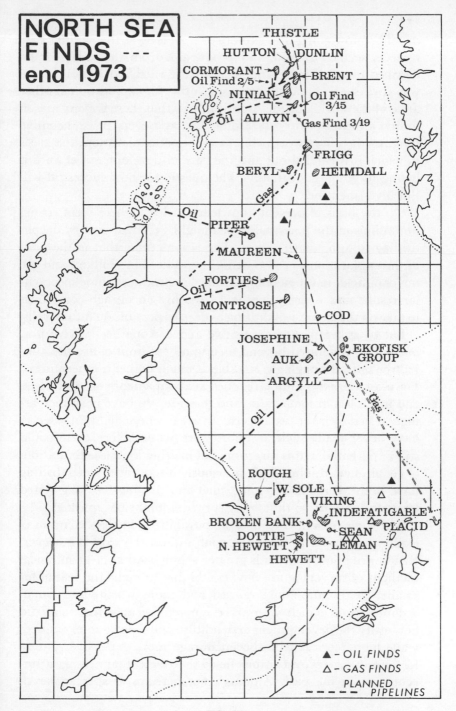

NORTH SEA FINDS --- end 1973

THISTLE
HUTTON · DUNLIN
CORMORANT ·
Oil Find 2/5 · BRENT
NINIAN
Oil Find 3/15
Oil ALWYN · Gas Find 3/19
FRIGG
BERYL · HEIMDALL
Gas
PIPER
MAUREEN
Oil
FORTIES
MONTROSE
COD
JOSEPHINE
AUK · EKOFISK GROUP
ARGYLL
Oil Gas
ROUGH
W. SOLE
VIKING
INDEFATIGABLE
BROKEN BANK · PLACID
DOTTIE SEAN
N. HEWETT · LEMAN
HEWETT

▲ – OIL FINDS
△ – GAS FINDS
PLANNED PIPELINES

FIG. 23

the effort now going into exploration. As long as the Middle East reserves seemed infinite and production from the area seemed destined to go on increasing at dramatic rates indefinitely, there was always a cost and supply disincentive for exploration in other areas of the world outside America. Now that there is every reason to find alternatives to the Middle East and every price incentive to make the effort worthwhile, the rate of exploration, the areas in which it is carried out and the pace of development of finds is bound to go up enormously. The question is how successful will this investment be.

On the basis of present knowledge at least, there seem strong limitations in the potential which might be uncovered by deeper drilling into the earth's crust. This is something that in the past, in the exploration of the Groningen gas field in Holland and the major oilfields in Libya, for example, has produced some spectacular results and rising prices will certainly encourage companies to take on the expense of drilling as deep as possible on future wells.

But oil is temperature sensitive and can neither survive nor, indeed, be generated beyond a certain heat limit of around 500° Fahrenheit (although gas is stable at much higher temperatures). For most areas this limit is reached at about a depth of 15,000 feet, and much less in some cases, and these depths have been increasingly tested over the last decade. In areas where plentiful reserves have never made such an effort worthwhile before there could still be potential in this direction, but in other areas the limitations of geophysical results which previously discouraged deep drilling have been overcome for some time and drilling has long tested lower formations, so that fewer surprises might be expected.

There do remain considerable possibilities in the extension of exploration in geographical spread around the world, however. Rising prices coupled with greater sophistication of geophysical methods of surveying are now combining to make the search for smaller fields both more refined and more worthwhile. Small accumulations which were once ignored because their size did not justify the development expenditure will now be re-examined, while higher prices will also encourage more expensive systems for improving recovery from fields that were again thought uneconomic in the past. Although not perhaps as spectacular in their results as some have claimed, these methods of improving recovery will serve to lengthen the period at which oilfields can sustain a high rate of output and raise the amounts that are ultimately recoverable.

High prices will also intensify the trend, already growing in the 1960s, for the oil search to extend to regions deep inland, or in jungle or arctic conditions, where the high costs of development previously seemed prohibitive. The most famous illustration of this is Alaska, where the discovery of a giant field on the North Slope by Atlantic Richfield–Humble (Exxon) and British Petroleum set off one of the most massive exploration efforts and largest auctions of licence areas ever recorded in the oil industry's history. Since then the continuous, and equally dramatic, delays in getting the Trans-Alaska pipeline built to take the oil to the warm-water port of Valdez has greatly reduced the rate of exploration. With the political go-ahead for the line, this is reviving rapidly and there are many in the industry who believe that the 10,000m. barrels of recoverable oil estimated in the area in 1970 could eventually be more than doubled.

The Alaska discoveries have in turn promoted new interest in the Canadian Arctic area, where conditions, if anything, are even more difficult than those in North Alaska. Early drilling produced a number of important gas finds but, up until 1973, the area was still far from offering up the oil promise that it had been widely held to contain. Outside of the Arctic regions, most on-shore areas in the West have been at least partially drilled and fairly extensively surveyed, so that the chances of really major discoveries seem rather less. But there may be considerable potential, only now beginning to be touched, in the Amazonian Basin and the jungles of South America.

China, too, is relatively unexplored. And, of course, there remains one of the largest prospective basins of any still left untouched in northern Siberia, where the extremes of climate, the distance from the market and the problems of shifting ice all present unparalleled problems of exploitation, quite aside from the political questions that any exploration and development of this region pose.

The really important development in recent exploration, and

the one that is likely to do most to change the face of the world's oil map, however, must be the move to off-shore exploration and development. The Continental Shelves surrounding the land masses of the earth contain very similar prospects to those on land being usually simply an extension of the same geological conditions. In some cases, as in the North Sea and around parts of the African coast, they can be much more attractive than on the neighbouring land because of specific geological occurrences. In other cases, such as off-shore Holland and off-shore Kuwait, they can prove relatively disappointing.

That there were strong prospects on the Continental Shelf has long been known. But, as with so much in the oil industry, technology and political will has tended to develop alongside rather than ahead of economic incentive, so that what started off as a fairly gradual movement of simply extending wells to the off-shore on piers where fields ran across the shore-line, has now become the most active area of all of exploration investment. Once again it has been in the United States where the main incentive to develop off-shore technology has come and it has been in the United States' Gulf that most of the techniques of drilling from mobile rigs and producing from fixed platforms have come from and where much of off-shore production has so far been developed since the war.

In the last ten years this development has spread to almost every Continental Shelf in the Eastern and Western Hemisphere and off-shore output is now taking a rapidly-rising share of world output outside America as well as inside it. According to a recent report by the United Nations, more than 100,000m. barrels of recoverable oil had been discovered off-shore by 1972, or a little less than 20 per cent of the total world proven reserves, while off-shore production in that year amounted to about 9m. barrels per day, or 18 per cent of the world total. Altogether off-shore exploration is now taking place off the coasts of a hundred countries, and production has commenced or is planned in about forty of them.

The most dramatic, and in some ways the most unexpected, development in this move has been the discovery of a series of 'giant' oilfields and several of the most prolific oil basins anywhere in the world outside the Middle East in the North Sea off the north-west coast of Europe. The story is an almost classic one of the surprises that can occur in exploration and the speed and scale on which the industry moves once a new 'play' begins. That oil companies should long have taken little interest in the area is not surprising. Decades of continual search from the First World War onwards on the lands surrounding the North Sea had produced little but a few small fields of oil and gas in Britain, Holland and Germany. The techniques of geophysical surveying in the sea had not yet been developed and the costs seemed prohibitive.

By the beginning of the 1960s, however, everything suddenly came together to change this attitude. In 1959 the Shell–Esso partnership shook the world with the discovery of an immense new gas field at Groningen near the Dutch coast in a situation that was similar to the smaller gas fields that had recently been established near the east coast of England. The obvious implication was that the two could be part of the same geological trend stretching beneath the southern part of the North Sea and this seemed further confirmed with BP and Shell, backed by Esso, used the seismic techniques then available to map the North Sea basins. At almost the same time an agreement was reached in Geneva in 1958 which set down the principles on which national governments could claim, and divide between them where their interests overlapped, the rights to minerals and other riches in the seabed out to a distance of 200 metres – or the edge of the Continental Shelf. And this enabled the various countries of the North Sea to start allocating their areas and negotiating boundaries early in the 1960s.

In 1962, Denmark awarded its off-shore concessions to a Danish shipping group. In 1964, Britain and Germany followed – Germany by offering the whole of her territory in one concession and Britain by dividing her territory into relatively small blocks of 100 square miles each and offering them to companies with favour given to those which had contributed to the British economy. A total of sixty-one companies applied for nearly 400 of the 960 blocks on offer and within three months of the licences being awarded to fifty-one of the companies, the first well was 'spudded in' on Boxing Day 1964.

Like a number which followed, it was unsuccessful. But the next year saw British Petroleum on its first well strike lucky with the West Sole Field, while 1966 saw Shell–Esso on its third well with the Leman Bank, one of the largest off-shore gas fields in the world and alone enough to supply more than double the amount of gas then being consumed in Britain. A year later, four major gas fields had been established and the gas industry of the United Kingdom had taken the momentous decision to convert its entire system to natural gas and to market the fuel on the basis that supplies could reach 4,000m. cubic feet per day by 1975, more than quadrupling the size of its sales in the process. The forecast has since proved remarkably accurate.

Even then, however, there were still doubts about the oil prospects, although a few small oil shows did turn up during the drilling. The gas found came from coal beds, which were unlikely to generate oil, and there was little in the drilling results made in the southern sector of the North Sea to suggest that conditions were favourable for large accumulations of oil. Nevertheless, a number of companies, notably Shell–Esso and Total in the United Kingdom sector and Phillips in the Norwegian sector, went on plugging away farther north with wells that cost £1m.– £2m. apiece and which, for several years, produced nothing but 'sniffs' of hydrocarbons. The industry was almost ready to give up entirely – and several companies had in fact decided to do just that – when Phillips finally came out with the news of a giant oilfield discovery, the Ekofisk Field, off the south-west coast of Norway in 1970.

Once again the industry geared itself up to a new play. The centre of activities was moved from Great Yarmouth in England to Aberdeen in Scotland. Large rigs capable of drilling in deep water were brought over in increasing numbers from other parts of the world. By the end of the year British Petroleum had once again made the first big find off the United Kingdom with its discovery of the giant Forties Field north-east of Aberdeen, the largest off-shore field so far discovered in the North Sea. Little more than a year later Shell–Esso had quietly drilled its first successful well on the giant Brent Field north-east of the Shetlands, thus establishing yet another prolific basin, the East Shetlands Basin. By the end of 1973, after the loss of three rigs, the expenditure of some £500m. on exploration and about twice as much on development of both oil and gas, the drilling of some 460 exploration wells and more false starts than the industry cares to remember, the North Sea had been proved as a major new oil and gas province of world importance. Total recoverable oil reserves were estimated at some 14,000–15,000m. barrels of crude oil, enough to sustain a rate of production in Norway that was potentially three times that country's total energy needs and enough to make the United Kingdom self-sufficient in oil as well as gas by the end of the decade. The importance to the surrounding countries could hardly be overestimated. The importance to Europe as the first really major oil source of its own, although still far short of fulfilling all its potential needs, was hardly less.

If the North Sea has shown that the distribution of the world's oil reserves can be altered and that really major new discoveries can be made, however, it is still far from certain how far this will prove true of other drilling. Exploration off-shore, which may now account for as much as two-thirds of the industry's exploration investment, is now covering most of the obvious prospective areas of the Continental Shelf and is certainly showing a wide-spread rate of success in South America, off the coast of West Africa, in South-East Asia and even in the Mediterranean, where finds have now been made not only off North Africa but also off Greece and Spain.

But nothing approaching the importance of the North Sea – where ultimately recoverable reserves are expected to total 30,000–40,000m. barrels or more, the equivalent of total current proven reserves in the United States – has yet been found and, as might be expected, much of the reserves so far discovered lie off the major traditional producing countries. Of the 115,000m. barrels of recoverable oil proved off-shore, as presented in the United Nations report of 1973, about 15 per cent were off-shore North America and over half were off-shore the Middle East, with Saudi Arabian off-shore reserves alone totalling some 40,000m. barrels of recoverable oil.

This picture may change with further drilling, particularly as it goes deeper out on to the shelves. Preliminary surveys have suggested that the Atlantic area off the east coast of North America especially could hold considerable reserves. Drilling off South-East Asia, which has long been held up by political problems, is now beginning to gather pace with considerable promise in the South China Sea and off Indonesia and Malaysia. Much of the off-shore territories of Russia are hardly touched, of course, and the vast area of the Continental Shelf off Northern Siberia could hold really immense reserves.

In addition to this, drilling off the north-west coast of Australia has already proved up some major gas reserves and considerable prospects for oil. The promising off-shore basins around Greenland and off northern Norway have yet to be tackled, while even fairly small discoveries of oil in areas such as the Mediterranean, parts of Africa and Latin America could make a considerable difference to local energy conditions. Overall, the proportion of off-shore oil production to total output seems almost certain to increase. By 1975, with all the development programmes now in motion, it could well total some 18–20m. barrels per day, or over a quarter of world output. By 1980, the proportion could be as much as half with a possible output of 40m. barrels per day.

Beyond that, the potential really depends on the very deep water off the edge of the Continental Shelf and on the ocean seabed itself. Prospects in this direction remain very uncertain. The equipment to drill in deep water of 1,000 feet and more already exists and the technology for producing from such depths is now being developed with research into methods for subsea completion, production and distribution of oil which can avoid the need for installing platforms to hold the producing equipment above the water altogether.

The Continental Slopes or Rises which extend from the edges of the Continental Shelf down to the ocean floor at 10,000 feet are certainly promising in the sense that they contain the same kind of mass of sediments that exist on land and there are many in the industry who feel that the prospects here could dramatically alter future oil prospects. But there are some misgivings about whether the right reservoirs exist in deeper water because of the nature of compaction of rock at great depths and the tendency of sand and porous limestone to accumulate near the coast rather than far from it. There are also fears that the sliding of masses of sediments which occurs down the slopes could have disturbed the processes of building up source and reservoir rocks.

Even greater misgivings have been expressed about the prospects on the ocean bed itself. Although a series of scientific wells has been drilled at various points of the ocean (which covers some three-quarters of the earth's surface) under a programme organized by the United States' oceanographical institute, and these have encountered natural gas in various parts, the geology of the rocks is fairly young, the thickness of sedimentary sequences is often thin and, where drilling has penetrated the ocean bed, the type of rock has been found generally non-prospective for oil in any great quantity.

Added to these doubts there are also considerable economic and political problems associated with deep-water drilling. The cost of production from water of several thousand feet or more, as at present estimated, could put oil from these sources at even higher prices than oil production from tar sands and shale. International agreements have yet to be reached on the thorny question of ownership of oil beyond the 200-metre depth and, while individual countries extend their boundaries farther and farther out, it remains a hotly-debated issue whether the ocean floors should be managed internationally or nationally.

Only actual drilling will tell whether the off-shore really does hold the solution to the world's future oil problems. The exploration industry is by nature an optimistic one but it is also one that has had to come increasingly to terms with the finite limits of potential oil sources and the difficulties of ever finding an oil province quite so productive as the Middle East, with its unique combination of immense sequence of reservoir rocks, well-formed structural traps and immensely productive source rocks of shale and even limestone.

If the North Sea and the Alaskan discoveries of the last decade have shown that major new oil provinces can still be discovered, they have also shown the tremendous costs that are now involved in exploration, the long lead-times between discovery and production that must now be faced and, not least, the intensifying political difficulties that new discoveries arouse. In Alaska, it

has been the environmental objections that have raised so powerful a pressure. Elsewhere it has been the governmental forces
which have been incited to action. In Australia, the government
has now moved to prevent export of off-shore gas, to take a
greater state participation in new exploration and to force greater
federal control over the states. In the North Sea, the Norwegian
government has acted to restrict the rate of development to suit
more closely the country's own economic ability to absorb the
revenues and social frictions arising from the off-shore finds. Even
in the United Kingdom, once regarded as the most secure of all
political environments for the oil industry, the forces of Scottish
and Welsh nationalism have been fired by oil, while the political
cry has been raised for higher tax terms, greater state control and
participation and even a slow-down of the rate of development
to suit local needs.

These forces are ones that the oil industry has always had to
live with. If they do not make the course of exploration and production any easier, they are unlikely to prevent it altogether. The
lure of the elusive 'black gold' that oil has been called is a powerful
one and its pursuit is likely to continue to form a major part of
industry activity for many years to come, whatever the degree of
success.

CHAPTER 29

Natural gas

'Natural gas . . . is without doubt *the* premium fossil
fuel of today.'

Dr Paul Tucker of Phillips Petroleum

While oil has proved the most reliable and most flexible of all
the sources of energy over the past generation, its sister fuel,
natural gas, has so far enjoyed a far more restricted pattern of
growth and in many ways a more abused career.

This is a waste and has long been recognized as such. A mixture
of methane and light hydrocarbon fractions, natural gas is in
many ways the most attractive of fuels. It is clean, versatile and
easily controllable in use. Save for transport (and even here tests
are being carried out using methanol and liquefied natural gas
as automotive fuel), it can fulfil nearly all the functions of oil,
coal and electricity from the generation of electricity in power
stations to providing fuel for domestic heating and light for houses
and streets. Like electricity, it can be turned on and off at will,
with the added advantages that it can be closely controlled at
the point of use and it can be stored, making it especially suited
to the seasonal requirements of space heating as well as specialized
manufacturing processes such as glass and pottery production.

Like oil, it can also be used as the raw material for a wide range
of chemical products. Methane, the main constituent of natural
gas, can provide a basic source for ammonia production and
hence fertilizers, as well as methanol and therefore plastics, and
carbon black, hence synthetic rubber. Ethane, butane and propane, the light hydrocarbon fractions present in gas, can be used
in the manufacture of ethylene and the other basic 'building
blocks' of the petrochemical industry, while other fractions such
as nitrogen have considerable value in the same field.

In terms of potential supplies, it has been found in widespread and often substantial quantities across the world, largely in association with and generated in the same way as oil. This 'associated gas', which is always present in solution with oil and often in sufficient quantity to form a 'gas cap' above it, forms an essential 'drive' mechanism to maintain oil pressures in fields. While in the past it has sometimes been regarded as a nuisance in actual production, requiring special treatment to prevent explosions, it can also form a sizable source of energy production in its own right, provided that a market can be found.

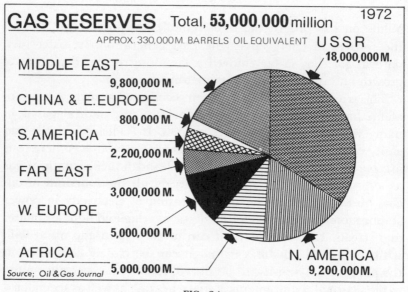

FIG. 24

Gas is also found in large accumulations on its own, either because its stability at higher temperatures than oil has allowed it to survive where oil has not or because it has been generated – as it sometimes is – from a different source than oil. Methane, for example, is often present in coal seams, the product of degeneration of freshwater vegetable matter, and it is from carboniferous coal measures rather than oil source rocks that the natural gas in the southern sector of the North Sea and in some other parts of the world is thought to have been generated.

There is evidence, too, that gas can be formed by inorganic chemical reaction. Methane gas is an important component in the atmosphere of planets of the solar system and the tails of comets are largely composed of solid methane. Considerable research is now being carried out into the possibilities of an inorganic source for methane on earth, and this has yielded some promising results, although there remains the problem that methane found at great depths and pressures may occur in the form of hydrates, which are uneconomic to produce.

Altogether, proven reserves of both associated and non-associated gas at the end of 1972 totalled some 53 trillion cubic metres, of which over a third lay in Russia, and over 9 trillion cubic metres were held by both the United States and Middle East. This is roughly equivalent, on a thermal basis, to about 330,000m. barrels of oil, or over half the total proven oil reserves in the world. As exploration continues and gas takes on more interest as a target, many in the industry believe that as much as twice this amount remains still to be discovered and that reserves of gas will ultimately prove roughly equivalent to those of oil.

In the markets where it has become available, it has become the fastest growing of all fuels. In the United States, where it has long been a favoured source of energy ever since the 1930s, it supplies almost 33 per cent of the country's total energy requirements and accounts for 50 per cent of the stationary energy market and the bulk of the domestic markets.

In the USSR it now accounts for nearly 25 per cent of the country's primary energy requirements, having shot up from less than 8 per cent in 1960. In Western Europe, the growth rate has been even faster, with consumption increasing from around 25,000m. cubic metres in 1966, or some 2·5 per cent of total primary energy demand, to about 135,000m. cubic metres, or around 10 per cent of the total. And this proportion is expected to grow to as much as 20 per cent by the end of the decade.

And yet, for all the obvious attractions of natural gas as an energy source and despite the wealth of reserves around the world, it has never taken off as an *international* energy commodity in the same way as oil. Compared to a reserve position that some put as high as two-thirds of that of oil, gas supplies only around 17 per cent of the world primary energy market compared to oil's 47 per cent share. If the USSR, Eastern Europe and China are excluded, the disparity is even greater, with gas accounting for some 19 per cent of total energy demand against oil's 53 per cent.

Although its position in some countries is well-established, in

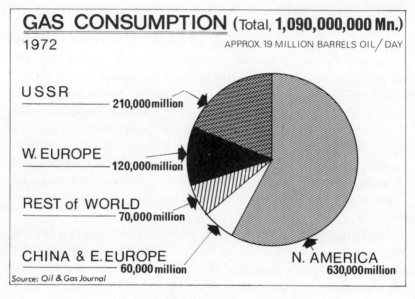

FIG. 25

other major energy markets such as Japan it has hardly got off the ground. While in countries where both the market and reserves are present, development has proceeded at an extremely rapid pace, in other areas with high reserves but without a near-by market, the reserves have hardly been exploited at all. Steps are now being taken to prevent some of the wastage previously involved with the flaring-off of associated gas in the major producing countries of the world, but as much as 35 per cent of Venezuelan production and 60 per cent of Iran's total production still goes up in smoke every year while large reserves of non-associated gas fields in Asia, the Middle East and elsewhere lie unexploited.

The limiting factor has been basically one of transport and distribution. Although natural gas is searched for and produced in the same way as oil, it suffers from a number of disadvantages in being taken from place to place. With a much lower thermal density than oil, it requires about five times the size of vessel to carry the same quantity of gas in terms of heat content as oil. Being a gaseous rather than liquid substance, it is harder to handle, more dangerous because it cannot be seen, and it has traditionally only been possible to transport it economically by pipeline.

Gas development has tended therefore to have had to wait on the development of pipeline technology. In China, 2,000 years ago, it was moved in bamboo pipes. In New York in the early nineteenth century it was transported in wooden pipes. The use of cast iron and lead pipes made the task easier as gas, then manufactured from the plentiful coal supplies of the times, developed as a primary lighting fuel in many cities of the last century. But without the materials and technology for 'pressurizing' the gas flows, and thus increasing both the amount that could go through the pipeline and the length it could travel, gas could only be carried over short distances and its development was limited to individual urban systems based on coal conversion rather than an extensive use of natural gas reserves, which were usually found at some distance from the major industrial markets.

The whole picture was changed in the 1920s when the development of acetylene welding and higher quality steels made possible the construction of large-diameter, thin-walled pipes capable of carrying gas under high pressures.

Inevitably it was in the United States that this breakthrough made the most immediate impact. In 1931 the first 1,000-mile pipeline was completed and gas from the major fields in the south-west started to be distributed through the country. Despite sometimes bitter opposition from coal interests and oil companies without large gas reserves, the penetration of gas into the market shot forwards after the war, encouraged by the lack of restriction on production and cheap prices. Development in Russia, the other major market with large reserves of its own, was rather slower due to the lack both of technology as well as steel materials and financial resources. But aided partly by deals to sell gas to the West in exchange for pipe, the 1950s and 1960s saw in the USSR too a dramatic push and by 1975 the country plans to have some 160,000 miles of gas pipeline system in operation.

In Europe, development has had to await the establishment of large indigenous reserves of natural gas. Discoveries during the war and in the immediate post-war period in France, Italy and Austria did enable regional progress to be made and, indeed, it was the gas discoveries in the Po Valley, the 'bomba petrolifere di Cortemaggiore', in 1949 that helped establish the legend of Enrico Mattei of ENI. But these finds, despite the enthusiasm with which they were greeted at the time, were never able to contribute more than a small proportion of the energy require-ments of the countries concerned, except in the smaller market of Austria, and it was not until Shell–Esso's unexpected and enormous discovery at Groningen in Holland, right by the major industrial markets of north-west Europe, that gas came into the European picture as a major fuel in its own right.

The transportation problem of gas has not only given the fuel a more restricted international role in the past compared to oil, it has also given the gas industry a very different marketing approach and structure. Where oil is basically a very flexible commodity in its distribution pattern, natural gas is relatively rigid in its source of supply and economics. Once a pipeline is laid, the market is almost irrevocably tied to that source and the industry has to suit its marketing policy to the rate at which its supplies are delivered and the limitations of bringing on new supplies until a further pipeline is laid.

Of necessity, therefore, the gas industry is one in which long-term planning and periods of very high capital investment are paramount and in which the marketing of gas, the balancing of seasonal and fixed loads and the relationship between bulk in-dustrial markets with few customers and the domestic market with many small consumers, have to be worked out against the limitation of pipeline capacity and the restrictions of basic supply. And this, coupled with the traditional monopoly position held in urban centres by the original gas manufacturing companies, has promoted large-scale, usually state or municipally-owned organizations for the transmission and sale of gas. Oil companies remain the basic producers of the great bulk of natural gas coming into the market but only rarely are they involved in the transmission of this gas to the market or the distribution and sale of the gas within the market.

In the United States, gas is bought from the producers by the major pipeline companies who then sell it to local municipal companies and major industrial users. In Europe, the structure is more varied, with the state taking a much larger share. In Britain, France and Italy, all gas sales and transmission are handled by state-controlled monopolies, while in Germany the local municipal companies still retain their independence. Europe is also different from the United States in that it was not until the 1960s that the oil companies became involved in gas production and distribution in a sizable way.

Thanks largely to Groningen and the North Sea, this is now changing. Shell and Esso in the NAM partnership in Holland are involved not only in the sale of gas in the Netherlands through a 40 per cent share with the State in Gasunie, but also in the export of gas outside the country. Partly through that, they also have shares in the Distrigaz monopoly concern, which transports and sells gas in Belgium, as well as in West German pipeline distribution. Shell alone now accounts for something like 20 per cent of all natural gas supplies in the Free World outside North America and the ranks of oil companies now producing gas in the Eastern Hemisphere are rapidly being swelled by the large gas finds off-shore being made by companies like Total and Phillips in the North Sea.

Nevertheless, the involvement of the oil companies in 'downstream' gas distribution remains at best limited to minority shareholdings in transmission companies and, for the most part, is confined to the role of the producer only selling to monopoly enterprises in each country. This in turn has put the oil companies in the past in the awkward position of selling a fuel which is highly competitive with oil to organizations which naturally want to use their monopoly-buying position to gain it at as cheap a price as possible in order to compete with them.

In the early years of North Sea exploration, the result was a bitter and long-drawn-out struggle between the producers of North Sea gas and the Gas Council (now called the British Gas Corporation) over the price to be paid for gas, and it was not until two years of negotiation had gone on and the companies had already invested in the development and subsea pipeline systems that the issue was finally settled in agreements and at a price which the companies never quite forgot or forgave. And this problem had been again repeated more recently in Holland in a struggle over the Placid group's right to sell its North Sea find directly to Germany or, under compulsion, to Gasunie in Holland.

This structural picture in the gas industry is unlikely to change. Indeed, government control of gas distribution and sales as a 'utility' is likely to increase in an age of serious energy concern. But what is changing rapidly is the attitude towards gas as a central factor in any future energy equation. During the great period of expansion of gas sales in the United States and Europe, in the 1960s, the market for the fuel was bought with low prices in order to ensure the maximum build-up possible. Gas was sold not just in competition to electricity and heating oils in the heating and quality-processing markets, where its high quality makes it an ideal fuel, but in replacement for residual fuel oil and coal used for bulk steam-raising in factories and in power stations, where much of it is wasted by the inefficiency of the conversion processes. In the United States, gas now takes up just about 50 per cent of the total industrial market for primary energy, compared with nearly 30 per cent for oil and around 20 per cent for coal and, more surprising, nearly 25 per cent of the electricity generation market.

Outside the United States the story, while not as extreme, has tended to be of the same pattern. The need to build up rapid sales has encouraged gas companies to sell too much too quickly into the bulk industrial and power station markets. The inevitable result has been that natural gas, which has taken so long to achieve a real impact on the energy markets, has now been sold to the point of severe strain on easily-available supplies. The ratio of reserves to production in the United States has fallen from 16:1 in 1966 to less than 11:1 by 1973. While gas from Alaska could provide one new source over the longer term, it is clear that the country is entering a period when indigenous supplies will fall further and further behind the potential growth in demand and will in all probability decline in absolute as well as relative terms. In Western Europe, the rapid surge of supplies from Groningen, the North Sea and elsewhere has itself created a demand which indigenous supplies cannot hope to sustain. The expected producing rates of the Groningen Field, over half of whose planned production has been committed on long-term export contracts, have now had to be revised sharply downwards in the light of actual production experience. Low prices have discouraged gas exploration in the southern sector of the United Kingdom. North Sea and the major new finds of associated and non-associated gas in the northern part of the North Sea will take some years to develop.

One thing that will occur in consequence – and none too late – is a re-examination of the place that gas ought to hold in the energy market and a growing conviction by governments of the need to ensure that it is sold as far as possible to the markets where its special values are best used. Already gas supply contracts are being increasingly tied to the price of lighter oil products and this, alongside stronger pollution controls, will tend to drive it to the highest value markets in heating and processing work.

At the same time there is an increasing effort now being undertaken to bring gas to the major industrialized markets of the West from a much greater variety of international sources. The dissipation of the Cold War in Europe has made possible the planning of massive imports from Russia into Western Europe by pipeline on a scale that would have been unthinkable a generation ago. Total contracts agreed or being negotiated by Western European authorities with the Soviet Union in 1973 involved a total future supply of 2,000m. cubic feet per day, or around 20,000m. cubic metres a year, and further supplies are still being discussed. By the end of the decade, when these have built up, the Soviet Union could be supplying towards 10 per cent of Western Europe's total gas consumption. Added to this, there is now considerable research being put into the possibilities of building subsea pipelines to take gas from the major fields of Algeria to Italy and France and, while such a project poses daunting questions of water depths and seabed conditions, it could come to fruition at the end of this decade or early in the next one. In the same way, a number of studies have been made in North America surveying the possibilities of building a great gas network to take supplies from Alaska and the recent discoveries in the Canadian Arctic across Canada down into the United States.

Even more dramatic in its implications, particularly for the oil industry, has been the development of large-scale projects to ship gas in liquefied form from the main oil-producing countries to Japan, the United States and Britain. First developed in the early 1950s by the Union Stockyard and Transit Company of Chicago as a way of circumventing its natural gas pipeline suppliers, the idea of freezing natural gas to sub-zero temperatures (thus reducing it to 1/600th of its original volume) and shipping it in specially prepared tankers was taken up by Shell, Continental Oil and the British Gas Council, which received the first experimental shipment of LNG (Liquefied Natural Gas) from the United States in 1959 and organized the world's first commercial traffic in the commodity with a deal for Algerian gas that began shipments in 1964. Ironically, the discoveries of gas in the North Sea soon after made further development of the trade by Britain unnecessary. But for other countries the opportunities, despite the expense involved in processing and shipping the fuel, were of very considerable interest indeed. As a means of 'peak shaving' – storing gas in order to meet the peaks of seasonal and daily demand over and above quantities regularly delivered by pipeline – LNG had obvious attractions economically. As a means of enabling new sources of supply to be brought in, it was of obvious interest to both the United States and Europe while for pollution-conscious Japan, the new trade provided almost the only means of getting large quantities of gas over the future. By the beginning of the 1970s, Shell was starting up one of the largest LNG projects ever negotiated, covering the delivery of 65m. tons of LNG a year to Japan from its large Brunei gas reserves; Libya and Algeria were supplying LNG in growing quantities to Italy, France and Spain, while United States gas companies were negotiating for supplies of even greater quantities of gas from

Algeria, Nigeria and the Middle East. Altogether, in 1973, there were some thirty LNG projects under discussion round the world, involving a potential trade equivalent of around 20,000m. cubic feet of gas per day – five times the total size of the United Kingdom gas industry – to be carried in around a hundred tankers by the 1980s. And there were some in the industry who thought that the figures could eventually double by the mid-1980s.

Predicted patterns of LNG trade
(in '000m. cubic metres)

	United States	Western Europe	Japan	Total
Early 1970s	—	10	8	18
Mid to late 1980s	50–120	30–50	30–60	110–230

Whether the LNG trade will in fact grow at this pace remains to be seen. Despite the high expectations that have been constantly held out for it over the last ten years, the practical development of projects has never matched up with first hopes. Early plants have been dogged by the problems of scaling up initial prototypes and, in both Algeria and Libya, have suffered continual breakdowns. Political problems have arisen at both the producing and consuming ends. Not least, the expense of the projects has been a powerful disincentive to rapid progress. Total investment costs in an LNG system average around $1.5m. for each million cubic feet per day of capacity, of which about 35 per cent is in the liquefaction and regasification plant at either end and about 65 per cent is in the cost of the tankers. The farther the distance to be travelled, the greater the cost. During the 1960s at least the delivered cost of LNG, at around $1.20 per million BTU (about 1,000 cubic feet of gas) on the east coast of the United States for example, threatened to be as much as four times the price of indigenous gas prevailing in the market.

The rapid escalation of all energy costs in the last few years has done much to overcome this obstacle and LNG must now be nearly competitive with higher grades of oil. There has also been some argument that the conversion of gas into methanol, which can be carried at ambient temperatures in tankers costing little more than an ordinary oil-product tanker, and which can be either converted back into natural gas at the other end or used directly as a chemical feedstock or clean fuel, could greatly reduce costs and bring added flexibility to the trade.

But the methanol route suffers at present from the extremely high cost of the initial conversion plant – as much as 75 per cent more than an equivalent liquefaction plant – and the comparatively low overall thermal efficiency of the conversion process, which requires as much as 50 per cent more natural gas feedstock than a comparable liquefaction operation. Under current technology at least, a methanol scheme would require very large throughputs and involve very long transportation distances before it could overcome these cost disadvantages.

The really major difficulty over the development of international tanker transport of gas, either in the form of methanol or LNG, however, remains essentially a political one. As with oil, the major industrialized countries would have to look mainly to the Middle East and the central producing countries for great increases in their gas supplies and this must raise questions about the basic price of gas and the political stability of supply. Unlike the oil trade, on the other hand, international transport of gas by tanker remains a rigid form of supply. Because of the high initial capital cost, contracts by their nature have to be long-term, tying the supplier and consumer to fifteen- or twenty-year periods in a set pattern of trade and, once the investment has been made, it is difficult to see the consumer being able to shop around for alternatives in a competitive trade should something go wrong.

Under these conditions there could be limitations to how far the consumer will be prepared to tie himself to gas imports in this way. In Europe, Germany has already set a limit to which it will allow Russian imports to take up its gas markets and a similar philosophy may also be applied to other countries in their policy towards LNG imports from North Africa and elsewhere.

In the United States, some of the LNG projects discussed at the beginning of the 1970s now appear to be going sour and, over the longer term, the country could well prefer to put more effort into producing substitute natural gas from coal or oil. Although more expensive, it is politically more secure. Equally in Japan, where a well-developed commercial and domestic market for gas has not yet been developed in any case, gas imports may seem only a partial answer to problems.

This will not prevent a substantial growth in international movements of gas by pipeline and tanker over the next decade. The energy needs of the world and the attractiveness of gas as a fuel are too strong for there not to be. The days when gas was greeted as an embarrassment to the oil industry are clearly in the past, even if the problems of developing it are not.

CHAPTER 30

Energy corporations?

'Oil companies must widen their interests in the energy
field if they are to remain in business. If they don't they
are going to die on the vine.'

*John Logan of Universal Oil Products**

If profit has been one of the driving forces of the oil industry,
the instinct for long-term survival and growth has been another.
As the foundation of the oil industry, the international integrated
oil companies have had to look at their role in terms of the future
trends and problems of the trade as a whole as well as the particular
ups and downs of demand–supply and prices in any one year. It
is this that has given them a quite different philosophy towards
pricing policy in the markets than smaller companies building
up sales in any one sector of the business and it is this that has
acted as a restraining influence in their actions in national
countries for fear of the long-term consequences of short-term
exploitation of opportunity. It is this, too, that has made the
industry increasingly look at opportunities outside the strict confines of oil trade in order to preserve their position in the changes
in direction that the oil business might ultimately take.

In the past the impetus towards diversification of investment
has arisen as much from economics as strategy. The *raison d'être*
of the international companies is in any case the diversification
of supply sources, transport facilities and markets and as the
scale of this trade has grown and as the profits in the bulk of it
declined during the 1960s, so the oil companies have inevitably
looked at ways of moving into the higher value-added areas of

* In remarks reported in the *Financial Times*, 2 April 1974.

chemicals and other fields in order to make the best possible use of their basic assets and managerial strengths.

In a business that touches on quite so many aspects of life, oil companies have become involved in a multitude of enterprises. Exxon, for example, expanded in the 1960s from the basic theme of motoring into the development of the whole field of leisure associated with transports, setting up a chain of motor hotels in Europe and was, at one time, looking to broader horizons of the holiday business as a whole. Shell, considering ways of building on its experience in oil drilling, has been actively examining the possibilities of using solution mining techniques in the exploration for evaporate minerals such as soluble potash and potassium magnesium-bearing salts. It, like other oil companies, has also gone into the field of metals production, both as an extension of its experience as a basic commodity trader and as a means of widening the range of targets involved in its far-flung exploration effort. Both Gulf Oil and Standard Oil Company of California (Socal), meanwhile, have built from their basic involvement in land an urban development business, constructing homes and civic centres. From circuses to car sales, there has been almost nothing that the oil companies have not at one time or another touched upon.

But most of these enterprises have been relatively small and insignificant compared to the size and investment involved in the industry's basic oil business and it has been on the areas most closely connected with their basic trade and experience – the products directly derived from oil and the development of other fuels in the energy field – where oil companies have tended to make the greatest impact.

The most important of these, and the business which has attracted by far the greatest investment from the oil industry, has been petrochemicals. Its potential value to the oil companies is obvious. The manufacture and sale of organic chemical materials based on oil or gas – synthetic rubbers, virtually the whole range of plastics, fibres, fertilizers, paints and even pharmaceuticals – has been one of the fastest growing of all businesses since the Second World War, achieving virtually double the rate of growth of inorganic chemical manufacture and helping to make the chemical industry as a whole one of the five largest in the world, whose production now threatens to equal and outstrip that of crude steel.

At the bulk production end, it shares many of the characteristics of the oil business, involving large-scale plant to achieve ever greater economies of scale; heavy capital investment in fixed costs; a similar technology of processing in many cases as well as an international trade by tanker and pipeline. In the initial stages of manufacturing the basic petrochemical materials from oil or gas, the oil companies not only have the advantage of secure raw material supply but also tighter integration of their chemical and oil refining operations. Even in volume terms, petrochemicals – which now account for some 5 per cent of all oil consumption – would be important to the oil industry while in terms of the added value they can give to oil products they are without rival.

Despite fierce competition from the established chemical companies, some of whom are as big as the oil companies in their own right, and despite arriving on the scene fairly late in the Eastern Hemisphere at least, the oil companies have expanded into petrochemicals at a rapid rate in the last decade.

Overall, the oil companies are estimated to have around 40 per cent of the total world investment made so far in petroleum-based chemical plants. The oil industry's gross investment in this field at the end of 1972 was estimated to have totalled nearly $16,000m., while its capital expenditures in this field over the preceding six years had averaged between $1,300m. to nearly $1,600m. per annum. Although this is still far from forming more than a small proportion of the industry's total investment and while the individual sales of the oil companies in this market are still far from rivalling those of the major chemical companies like ICI, Du Pont and Hoechst, it has still been enough to put two oil companies at least – Shell and Exxon – into the ranks of the top twenty world chemical companies.

Under currently foreseeable circumstances, it is difficult to envisage any other form of 'downstream' diversification by the oil companies achieving quite the same weight as petrochemicals, nor does it seem likely that the oil companies could take on anything on the same scale in view of the growing opportunities and complexity of business that chemical manufacture still provides. But one other more recent market which has opened up is the production of proteins from oil through a process of fermentation. Pioneered by British Petroleum, the process, which yields a dried yeast which can be used in animal feeds, could provide an important route to meeting the world deficit in protein production, currently estimated at nearly 25m. tons against a world production of 80m. tons annually.

Chemical sales of main oil companies in 1971

	$ millions
Royal Dutch–Shell	1,414
Exxon	1,077
Occidental	705
BP	650
Phillips Petroleum	495
Mobil Oil	420
Standard Oil (Indiana)	375
Texaco	370
Gulf Oil	360
Continental Oil	360

Sales of main chemical companies in 1971

	$ millions
ICI	3,886
Du Pont	3,848
Hoechst	3,428
BASF	3,131
Union Carbide	3,038
Montedison	2,723

Already BP, which discovered the process accidentally in France during the late 1950s while investigating ways of removing waxy compounds from oil, has two plants in production in Scotland and France which have been selling products since 1972 on the open market and is now building a full-scale, 100,000-ton-a-year manufacturing centre in partnership with ENI in Italy.

Shell meanwhile is working on a process for producing single-cell protein from natural gas while Exxon has long been in partnership with Nestlé following other routes. High oil prices and worries about shortages of basic fuel could do something to retard progress and development must still overcome some deep-seated suspicions among governments and public over the 'right-ness' of non-natural food products. But successive tests by various authorities round the world have generally approved the safety of at least BP's products for human and animal consumption, while the call on feedstock, at one ton of product for one ton of oil, is not great in comparison to total oil use in the world.

If diversification into petroleum-based chemicals and other products has proved a major area of oil industry investment in the past, when oil companies were seeking to expand the market constantly for their ever-increasing supplies of oil, 'lateral' diversification into other energy sources may now take on much greater force as oil supplies become strained and its position in the market begins to become vulnerable to other, and perhaps cheaper, fuels. For years oil companies, especially in the United States, have been quietly taking positions in coal, uranium, nuclear technology and synthetic oils. Now is the time when these moves should come to fruition and begin to make a substantial impact on total industry investment and sales.

The most logical step is probably the investment now being undertaken in 'unconventional' supplies of oil from tar sands and shale. Leases on the Canadian tar sands have been taken out by almost all the main oil companies, including BP, Socal, Home Oil, Union Oil of California and Standard Oil of Indiana (Amoco). A pilot plant has been operated by Sun Oil since 1968 and within the next ten years nearly a dozen plants of about 100,000 barrels per day capacity each are expected to be in operation or under construction, including projects already announced by Shell; the Syncrude consortium of Atlantic Rich-field, Imperial, Gulf and Cities Service; and a consortium headed by Petrofina next to Shell's leases.

In the United States, development is now going rapidly ahead with the auction of new federal oil shale tracts at which prices of several hundred million dollars were achieved. Most of the main oil companies, including Shell, Gulf and many of the larger independent companies, have bid, whilst all the majors and most of the independents are now involved in various experimental projects to try out various mining and *in situ* production techniques. At least one project, organized by a consortium of Atlantic Richfield, Ashaland Oil, Shell and Oil Shale Corporation, is now moving towards the construction of a commercial plant and, while it will take some years before recovery and production methods are fully tested and accepted, others are likely to follow now that the price of oil has made it commercially attractive.

Worries about the future of oil supplies has also taken the oil companies into other non-oil-related energy projects, particularly in the United States where protection of indigenous fuels and a free market in energy has induced the oil companies to broaden their coverage at a rapid pace in the last five to ten years. Coal exploration and production especially have seen a large-scale entry of the oil companies both for the alternative fuel it provides to oil and the possible source of synthetic hydrocarbons over the future. Largely through acquisition of coal companies facing financing problems during the hard years of low prices in the 1960s, companies like Continental Oil (which purchased Consolidation), Gulf (which bought the Pittsburgh and Midway Coal Company), Occidental Petroleum (which purchased the Island Creek Coal Company), Exxon and BP–Sohio have all gained substantial coal producing interests in the United States while most of the other main oil companies have taken interests in coal exploration in North America. More recently, Shell, which is now preparing to produce coal in Canada, has also signed deals for coal exploration and development outside America in Indonesia, southern Africa and Australia, while BP, which has been noticeably cautious about diversification away from oil in the past, has now set up a subsidiary to pursue similar schemes.

The provision of uranium and the development of nuclear technology have also received increasing attention from the oil companies. Conoco, Exxon, Gulf, Kerr-McGee and Sohio are all producing or planning to produce uranium in the United States and, here again, most of the other large companies are involved in prospecting one way or another. Some of the bigger companies, with large research establishments, are taking this further as well by involving themselves in uranium enrichment and nuclear power station design and construction.

The most ambitious company so far in this direction has been Gulf Oil, which in the last decade has taken a leading part not only in fuel reprocessing but in the design, manufacture and marketing of High Temperature Gas-Cooled Reactor (HTGR) systems, already in operation in pilot-plant form, as well as research into Gas-Cooled Fast Breeder Reactors (GCFRs) and controlled fusion. And in this it has been joined by Shell, which was already involved in a European gas centrifuge project for the enrichment of uranium, and which bought a 50 per cent stake in Gulf's nuclear interests in 1973. Other companies have been less ambitious in this direction, but Exxon at least has invested a considerable amount in research and development of particular technologies associated with nuclear power, including a programme, in association with General Electric Company, to develop commercial uranium enrichment plant and a programme with Avco Corporation to develop a new isotopes separation process based on laser technology.

Nor is oil shale and tar sand production, coal exploration and nuclear development the whole story as far as oil industry diversification is concerned. The prospects for the development of geothermal power, particularly in the Geysers area of northern California, has attracted Union Oil of California (which has the largest operating unit to produce electricity from this source in tne world), Signal, Socal, Phillips, Getty and Shell. Occidental is now building a plant to produce fuel oil from municipal refuse in California. Shell and Exxon are both researching actively into solar power and Shell has for some years been doing considerable work into fuel cells and nearly all the major companies have now become involved in research into ways of producing substitute natural gas (SNG) and synthetic oil from coal.

Altogether the oil industry now accounts for something like 50 per cent of the uranium reserves and some 40 per cent of the uranium milling capacity of the United States, as well as nearly 33 per cent of United States' coal reserves, most of the allocated leases on tar sands and oil shale in North America and much of the existing and potential geothermal power production in North America. And higher energy prices and increasing pressure to develop alternative supplies of energy are likely to intensify this tendency greatly over the future, both within and without North America.

Whether the move is a good or bad thing, and how far it will go, are different questions. Opinions within the oil industry are still divided on the subject and companies have tended to follow different policies. For some, diversification is at best a means of establishing a position for the long-term future but should not be allowed to dilute the investment and growth in what remains the the main business of the industry – oil and gas trade. For others, oil companies will have to turn themselves into energy corporations if they are to survive the future. Governments and commentators outside the industry, too, seem sometimes torn between the desire to see the oil industry's undoubted talents of management and fund raising on an international level give new impetus to the development of alternative energy forms, while others have become seriously worried, with equal reason, about the monopolistic tendencies that might result from the creation of monolithic energy enterprises suppressing the competition between fuels.

Despite the pressures undoubtedly pushing the oil industry into further diversification, however, there remain a number of reasons why the move should be less overwhelming than either the proponents of 'energy corporations' might hope or critics might fear. The great problem of diversification for any large-scale modern enterprise lies in the difficulty of moving into any field that can make a really substantial impact on their overall returns and sales. And this is particularly so of the oil industry in view of the sheer size of its traditional business. For all the investments made in petrochemicals, this field still accounts for no more than 6–7 per cent of the industry's total investment in fixed assets and about the same proportion of its current annual capital expenditure.

The fields where it can most logically enter are usually also the fields where it is likely to meet the greatest competition from existing large companies with greater experience and technology. The petrochemical market is perhaps the most classic example. The entry of the oil companies in the later 1950s and 1960s, coming at a time when the traditional chemical companies were themselves beginning to realize the advantages of economies of scale and were starting to invest heavily in new plant, served only to increase the surplus production on the market, reduce prices and depress returns. Far from providing an alternative source of returns to oil, it only added further problems. More recently conditions have improved as the low returns of the later 1960s led to falling invest-ment and hence shortages of supply. But, although the oil com-panies are now looking with renewed interest in the market as one of the most attractive long-term uses for oil, the lesson has prob-ably been learned Many of the major oil groups have already divested themselves of the least profitable parts of this business and in the future there appears to be a greater feeling of the need for closer co-operation between oil and chemical companies to prevent the cycles of surplus and shortage recurring, just as there appears to be a greater awareness of the dangers of pursuing integration of chemical operations too far down into end uses such as fibres and clothing, where the management problems are quite different and the competition from established companies even greater.

The same considerations must also apply to alternative energy forms. In the case of United States' coal, the entry of the oil companies was probably encouraged by the poor financial state of the coal industry in the 1960s. In other fields such as tar sand and shale oil production, the oil companies are clearly particu-larly well-suited to tackle the problems because of the direct relevance to the oil market. But in other fields such as nuclear power, the problems of technology and management are of quite a different order than those of oil. The structure of the industry is unique and the ground is already held by major companies such as Westinghouse and General Electric whose experience, research and investment will be hard to compete with. The costs are enormous, the lead-times before cash flows are achieved are horrendous and the risks if the system does not prove popular with the electricity generation authorities, are quite staggering even by oil industry standards. Save for a few exceptions perhaps, the oil industry as a whole therefore will probably prefer to remain as suppliers of the basic fuels rather than contractors and designers of the plant itself.

Political problems may also restrict the role of the oil companies in diversified energy supplies. Outside of North America, coal, like gas, is very largely in the hands of centralized state or state-supported monopoly concerns, while the mining of coal or other minerals and metals worldwide is already firmly established in the hands of the international mining companies and is just as vulnerable, if not more so, to all the political problems of nationalism and nationalization as oil has proved to be. Nuclear power and electricity generation are equally treated as public utilities in most Eastern Hemisphere countries, and thus provide serious drawbacks to any would-be seller of nuclear systems or fuel battling against monopoly markets. Even in America, where a free market does still operate, one would not be surprised to see restrictions placed on oil company diversification on monopoly grounds.

Ultimately, the issue also comes down to one of finance. Any energy programme today involves tremendous capital expenditure, long lead-times and even longer periods before any real cash flow is achieved. The need for oil has in no way lessened in the past few years, while its appetite for ever greater quantities of capital investment has only increased. Considering the scale of these financial requirements, the oil companies will constantly be forced to choose between projects before them. For some it will be worth the risk and the cost to diversify on a major scale. For many it will not.

Oil, for all the prophets of doom and forthcoming revolutions in energy technology, will be around for a long time to come and it will be many years before oil ceases to be the central business of the oil industry.

CHAPTER 31

Tides and tempests

'Really oil is almost a noble product. . . . Why finish
this noble product in, say, thirty years' time when
thousands of billions of tons of coal remain in the
ground.'

*The Shah of Iran**

Journalists are like academics in that they find the past and
present always entirely inevitable and the future always entirely
unpredictable. And this is so especially of the energy field, where
experience has really provided only one certain lesson – that the
surest way to get it wrong is to extrapolate from past trends. From
almost the beginning, planners and observers have tended to be
mistaken on some things all of the time and, quite often, on all
things at the same time.

A hundred years ago, some of the wisest brains were predicting
a certain shortage of coal in the world. Ten years ago, they were
predicting coal's almost certain demise in the face of the relent-
less march of oil down a road of ever-decreasing prices and ever-
increasing impact on the energy scene. In fact, both predictions
have proved off the mark, and when the first great dramas of the
early 1970s occurred with the Libyan price demands of 1970, it
was shown that almost every company and most governments
had seriously underestimated the growth in energy demand and
the ability of the oil industry to supply it. Yet a year later and the
pendulum had swung right back again to renewed surpluses and
falling prices, and companies which had based their investment
on the urgent need for tankers and refining capacity were once
more embarrassed by the opposite.

* Announcing a further doubling of oil prices to the world in Tehran,
December 1973.

Two years later, in 1975, everything was back to confusion again as the Arab producers made the one move that most experts had been led to believe would never happen – the concerted use of oil supply as a political weapon by conservative and radical oil states. Now, in early 1974, this too has proved exaggerated. The fears of impending disaster and collapse engendered by the actions of the producers have failed to materialize and industry and governments alike are torn between those who feel that the final warning bell has been sounded in man's unthinking pursuit of wealth based on the exploitation of the world's finite resources and those who believe that the seminal events of the oil industry during the early 1970s provide just a peak in a cyclical movement of investment and price that will bring its own balance soon enough.

In this uncertainty, it would be a brave man who drew too precise a conclusion from the current situation and still braver for a reporter whose primary efforts have been devoted to relaying the wisdom and worries of the time rather than piercing beyond them. Undoubtedly there have been some central trends in the oil industry that have been apparent for some time. The basic conflict between an international oil industry intent on maximizing the flexibility of its sources and markets and individual producing countries determined to harness the rate and manner of exploitation of their natural resources to their own national needs has been at the heart of the oil business ever since OPEC was founded a decade ago. As the pressure of growing world demand for oil increasingly concentrated the sources of supply for the consumer on a small group of oil-rich nations, so these nations were bound to seize control of prices and output as soon as the fundamental conditions of supply and demand swung in their favour.

In the same way it is possible to see in retrospect that the tremendously rapid shift of the industrialized world since the war towards oil as its primary source of energy was bound to cause severe strains on the system in view of the disparity between the price of oil and the price of alternative forms of energy. While it might have been perfectly reasonable to argue, as some did at the time, that the price of oil to the consumer was still too high in relationship to its cost of production, the really important

point was its relationship to the cost of alternative fuels. Unless oil
was virtually infinite in its reserves and virtually limitless in its
geographical spread – neither of which has been true – there was
bound to come a point where the sources of supply would become
strained and the price of hydrocarbons jump to the level of cost of
their alternatives at some time or another. In this sense, the Arab
producers merely brought forward the day by their actions in
1973. They did not artificially create it out of nothing.

Even accepting this, however, there has been much in the
developments of the past which has been the product of accident,
or at least coincidence, rather than the remorseless pressure of
predetermined economic trends. If the dramas of the early 1970s
have been the result of a marginal swing in supply from relative
surplus to relative shortage, human error and accident – the
delays in bringing forth oil from Alaska, the mistakes of government policy over regulated gas prices in the United States, the
pressure of the environmentalists, the psychology of a capital-
intensive industry increasingly shaped by the cyclical pattern of
investment and hence supply – must take their due share of
responsibility for that swing.

Despite the sometimes unfortunate appearance of the oil
industry as one conglomerate mass of uniform opinion and
direction, there have been wide and passionately held differences
in opinion about what should be done at any one time in as well
as outside the oil companies. It is a moot, if theoretical, point,
what would have happened if the vast reserves of the Middle East
had not been held so strongly by the majors in their interlocking
concessionaire groups and if the Gulf had seen the same kind of
competition for development that Libya and Nigeria were to in
the 1960s. It is an equally real, if academic, question to ask what
would have happened if the oil companies had taken a different,
more united stand against producer demands in Libya in 1970
or in Tehran and Tripoli in the following year or what might
have developed if the consumer governments had intervened
right from the start of negotiations with a strong and united stand
of their own. What would have been the result if there had been
no Libyan revolution in 1969, if President Sadat had not moved
across the Canal in October 1973, if Alaska had come in on

schedule, if the consumer governments had better protected their coal industries over the years of decline? Perhaps the answer would have been the same events with a different time schedule, perhaps not. But to ascribe all the developments in oil to fundamental and inevitable forces would be to rob them of the very real choice that companies and governments alike faced at the time, to underestimate seriously the genuine variety of opinions that emerged during the course of developments and to disregard the many avoidable decisions and the particular events which helped to shape the history of the industry.

It is this combination of economic logic, political intervention, particular accident and human fallibility which makes the future of oil and the oil business so difficult to predict in the wake of recent events. On the one hand, at a time when energy supply and demand are essentially commercial judgements about price and return, the unparalleled escalation of oil prices introduced by the oil states must have the effect of bringing its own equilibrium again. New oil sources, revised reserve estimates and alternative fuels will all come on the scene at a rate few would have predicted at the beginning of the decade as resources are diverted into this area at the same time as demand itself is depressed by the new costs of oil. With all the resources available in the world, both in fossil fuels and their replacements, there is no absolute reason why the world's economy should be brought to its knees for lack of supply. And if there is a short-term economic reason for immediate dislocation, prices alone may be enough to swing the situation back to plenty within a relatively short period. Certainly it is hard to see in the logic of affairs oil prices being able to sustain any level in real terms much beyond what will be needed to bring alternatives into play – a figure that some believe to be below the rates reached at the height of panic in 1973–74.

On the other hand, there are equally forces that could prevent the natural law of supply and demand working as efficiently or as fast as some suggest. The lead-times involved in bringing on alternatives to oil or even new oil sources and the sheer engineering and contracting effort that will be necessary to produce the volumes required, still present formidable obstacles to the development of new energy patterns in the West.

Until more secure patterns of energy are achieved by the consumer both in terms of the balance between fuels and between oil sources, the question of price and supply will always be intimately bound up with the attitudes of the major oil producers. By restricting or increasing production, the producing states, and countries such as Saudi Arabia in particular, can change the margins of surplus to shortage as quickly and as dramatically as they did in 1973. And the consumers' ability to respond to this will in turn be dependent on investment and on politics – themselves factors which are at the mercy of government attitudes, social feelings towards profit and environment, company planning and the unknowns of oil discovery and technological breakthrough.

This is not necessarily to dodge the issue. The opening years of the 1970s have seen the oil business go through a revolution. Like most revolutions, history may well show the developments both less 'new' and less radical than those who participated in them believed. But, whether the 'oil crisis' proves exaggerated or not and whether the events of 1970–74 are followed by a period of renewed growth or further dislocation, the implications of what has happened in the last few years can hardly be overestimated. After two decades of ceaseless and unquestioned growth, the world has been suddenly faced with the prospect that oil can no longer sustain the burden of energy demand that it has in the past, that its resources are finite, its cost has been undervalued and that the geographical distribution of its reserves are fraught with problems.

What has happened is not just a structural change in the ownership and direction of crude oil supplies but a fundamental shift in condition from relative resource abundance in the world to relative resource scarcity involving profound questions of the transfer of wealth from the consuming to the producing countries, the re-alignment of position between resource-rich and resource-poor nations and the re-orientation of traditional patterns of resource consumption not only of oil but of other basic commodities. In the oil field, as in other areas, this has entailed a basic re-think among consumers about their relations with the producing countries, about their involvement in the oil industry and about their role in the market. If energy can no longer be taken for granted, nor can the systems by which it has been historically handled.

This must affect not only the way in which oil continues to be used over the coming decades but also the structure of the industry itself. The message of the events of the early 1970s may well be that oil will reach its peak as the major fuel source of the developed and developing world within this decade and that thereafter its position in energy patterns will begin to decline. And as oil loses its predominance in the energy scene, so may the international oil companies lose their predominance in the oil scene.

And yet it is too early by far to hold a wake over the oil business. Oil has come to take on the importance in the world's economy which it has, not simply because it was there but because of all fuels it is the most easily handled, the cheapest and the most versatile in use. Man is using it not only to provide heat and electricity, but to fuel his transport, lubricate his machines and provide the raw material for the fertilizers in his fields, the clothing that he wears, the plastics which he moulds and the protein which he feeds to his animals.

So too with the industry. The dominance of the international oil companies may owe much to historical accident, to intrigue, manipulation and to the past realities of imperialism and geo-political power in the world. But it also owes much to the basic need in the trade for integrated and flexible supply systems, balancing out sources with markets and growth with investment. If the central problem of any international commodity as important as oil is to ensure the greatest degree of security of supply and stability of price against the vagaries of individual national actions, the uncertainties of the cyclical nature of industrial investment and the changes in the world patterns of supply and demand, then the international oil industry has managed to effect this with remarkable efficiency and surprising success over the past.

It is a role which has inevitably aroused friction between the national needs of consumers and producers and the international requirements of the business. It has often aroused the wrath of those who would prefer to see the maximum benefit of particular circumstances of supply and demand at any one time going to producer *or* consumer in the form of higher *or* lower prices. And it is a role which has been found to have all too obvious limitations during the climactic events of the early 1970s. But it is not a role for which there is any easy substitute in national government management or dismemberment of its integrated form, any more than it will be easy to find any substitute for the versatility and convenience that are oil's basic characteristics as a fuel.

Developments have taken much of the bloom of youth away from oil just as they have taken much of the glamour from the oil industry. During the coming years, both will be subject to increasing pressures and restrictions as the questions of internationalism versus nationalism grow more intense. But what developments have not taken away is the importance of oil in the world's economy and the importance of some kind of international system to supply it. Nor are they likely to for a long time to come.

Appendices

Select Bibliography

Index

Appendix 1

The big seven companies ('the majors') in comparison with some other
prominent international oil companies and well-known companies from
outside the oil industry.

The big seven

Name	Nation-ality	Sales $m.	Assets $m.	Profits $m.	Employees
Exxon	US	20,310	21,558	1,531	141,000
Royal Dutch–Shell	Dutch–British	14,060	20,067	705	174,000
Mobil Oil	US	9,166	9,217	574	75,400
Texaco	US	8,693	12,032	889	76,496
Gulf Oil	US	6,243	9,324	197	57,500
Standard Oil of California	US	5,829	8,084	547	41,497
British Petroleum	British	5,712	8,161	176	70,000

Other prominent oil companies

Name	Nation-ality	Sales $m.	Assets $m.	Profits $m.	Employees
Standard Oil of Indiana	US	4,503	6,186	375	46,627
Continental Oil	US	3,415	3,250	170	38,092
Atlantic Richfield	US	3,321	4,629	196	27,756
Phillips Petroleum	US	2,513	3,270	148	35,265
Occidental	US	2,487	2,562	10	31,500
Compagnie Française des Pétroles	French	2,806	3,926	114	24,000
ENI	Italian	2,748	7,088	18	78,918
Elf–ERAP	French	2,396	3,693	11	20,426
Petrofina	Belgian	1,566	2,090	70	23,300

Companies outside the oil industry

Name	Nationality	Sales $m.	Assets $m.	Profits $m.	Employees
General Motors	US	30,435	18,273	2,162	759,543
General Electric	US	10,239	7,402	530	369,000
Unilever	British–Dutch	8,864	4,681	332	337,000
Nippon Steel	Japanese	5,364	8,623	68	98,714
Volkswagenwerk	German	5,016	3,494	60	192,083
ICI	British	4,236	5,487	229	199,000
Nestlé	Swiss	4,130	3,264	170	116,034
Goodyear Tire and Rubber	US	4,072	3,477	193	145,201
Mitsubishi Heavy Industries	Japanese	3,980	7,264	59	110,563
British Steel	British	3,630	3,466	7	229,000
British Leyland	British	3,248	2,085	60	190,841
Dunlop Pirelli	British–Italian	2,746	2,793	9	170,000
British American Tobacco	British	2,570	2,901	197	110,000
National Coal Board	British	2,537	1,285	(−205)	268,000

Source: *Fortune*
All figures apply to 1972

Appendix 2

Conversion factors
(Approximate)

1 American barrel = 42 American gallons
  = 159 litres
  = 35 imperial gallons

American barrels per day multiplied by 50 = approx. tons a year.
Thus a production of 20,000 barrels a day = 1m. tons a year.

1 imperial gallon = 1·2 American gallons
  = 4·6 litres
  1 metre = 3·281 feet
  = 0·001 kilometres
  1 kilometre = 0·621 miles
  1 mile = 1·609 kilometres

1 metric ton of :	*American barrels*
Crude oil – light–medium	7·2
heavy	6·5
Aviation gasoline (petrol)	8·9
Motor gasoline (petrol)	8·6
Kerosene	7·8
Diesel oils and gas oils	7·3
Heavy fuel oil	6·7
Lubricating oils	6·9

Appendix 3

The main oil-producing areas

Area	Production (in m. barrels per day) 1973	1963	% of world total	Average annual increase 1963–73
North America	13·6	9·8	22·8	3·3
Caribbean	3·8	3·6	7·0	0·6
South America	0·9	0·5	1·7	6·9
Western Europe	0·5	0·4	0·7	1·7
Middle East	21·1	6·8	36·8	11·9
Africa	6·0	1·2	10·3	17·6
South-East Asia	1·6	0·5	2·9	11·7
USSR, E. Europe and China	9·6	4·6	16·8	7·5
Country				
US	10·9	8·6	18·3	2·3
USSR	8·5	4·1	14·8	7·4
Saudi Arabia	7·4	1·6	12·9	16·3
Iran	5·9	1·5	10·3	14·9
Venezuela	3·5	3·3	6·3	0·5
Kuwait	2·8	2·0	4·9	3·6
Libya	2·2	0·5	3·7	16·7
Canada	2·1	0·8	3·6	10·3
Nigeria	2·1	0·1	3·5	38·7
Iraq	2·0	1·2	3·4	5·5
Abu Dhabi	1·3	0·1	2·2	37·2
Indonesia	1·3	0·5	2·2	11·1
Algeria	1·1	1·5	1·8	7·9
World total	57·7	27·4	100	7·7

Appendix 4

The main oil-consuming areas

Area	Consumption (in m. barrels per day) 1973	1963	% of world total	Average annual increase 1963–73
Western Hemisphere	22·1	13·4	38·6	5·1
US	16·8	10·6	19·5	4·7
Canada	1·8	1·0	3·0	5·9
Western Europe	15·2	6·1	27·0	9·4
W. Germany	3·0	1·2	5·4	9·6
France	2·6	0·9	4·5	11·9
UK	2·3	1·2	4·1	6·2
Italy	2·1	0·8	3·8	10·0
Benelux	1·5	0·6	2·7	8·7
Scandinavia	1·1	0·6	2·0	7·0
Japan	5·4	1·2	9·7	15·6
USSR, E. Europe and China	8·8	3·9	15·7	
Total E. Hemisphere	34·4	13·5	61·4	9·7
World total	56·4	26·9	100	7·7

Appendix 5

World 'published proved' oil reserves at end 1973

	Barrels ('ooom.)	% of world total
US	41·8	6·3
Canada	9·3	1·4
Caribbean	17·6	2·9
Other Western Hemisphere	14·0	2·2
Total Western Hemisphere	82·7	12·8
Western Europe	16·4	2·6
Africa	67·3	10·4
Middle East	349·7	55·4
USSR, E. Europe and China	103·0	16·3
Other Eastern Hemisphere	15·6	2·5
Total Eastern Hemisphere	552·0	87·2
World total	634·7	100·0

Source for this and preceding tables: *BP statistical review of the world oil industry 1973*

Select Bibliography

Study in Power: John D. Rockefeller, Industrialist and Philanthropist, Allan Nevins (Charles Scribner's Sons).

Marcus Samuel, First Viscount Bearsted and Founder of Shell, Robert Henriques (Barrie and Rockliff).

The History of the Standard Oil Company, Ida M. Tarbell (McClure, Phillips & Co.).

Since Spindletop, Craig Thompson (Gulf Oil Corporation).

The Texaco Story, The first Fifty Years, 1902–52, Marquis James (The Texas Company).

Mr Five Per Cent: the Biography of Calouste Gulbenkian, Ralph Hewins (Hutchinson).

Adventure in Oil: the Story of British Petroleum, Henry Longhurst (Sidgwick & Jackson).

The World Crisis, 1911–14, Winston Churchill (Thornton Butterworth).

Report on the International Petroleum Cartel, US Federal Trade Commission, 1952.

Oil in the Middle East, S. H. Longrigg (Oxford University Press).

Foreign Oil and the Free World, Leonard M. Fanning (McGraw-Hill).

Pantaraxia, Nubar Gulbenkian (Hutchinson).

The Super-Americans, John Bainbridge (Doubleday & Co.).

Oil Companies and Governments, J. E. Hartshorn (Faber & Faber).

Freedom for Fuel, Georg Tugendhat (Institute of Economic Affairs).

Essentials of Petroleum, P. H. Frankel (Chapman & Hall).

Mattei: Oil and Power Politics, P. H. Frankel (Faber & Faber).

Oil: The Facts of Life, P. H. Frankel (Weidenfeld & Nicolson).

The Price of Middle East Oil, Wayne A. Leeman (Cornell University Press).

A Financial Analysis of Middle Eastern Oil Concessions: 1901–65, Zuhayr Mikdashi (Frederick A. Praeger).

Crude Oil Prices in the Middle East, Helmut J. Frank (Frederick A. Praeger).

Petroleum in Venezuela, Edwin Lieuwen (University of California).

Middle East Oil, George W. Stocking (Allen Lane, The Penguin Press).

Oil and World Power, Peter R. Odell (Penguin Books).

Power from the Sea, Clive Callow (Victor Gollancz).

Competition, Ltd.: The Marketing of Gasoline, Fred C. Allvine and James M. Patterson (Indiana University Press).

Continuity and Change in the World Oil Industry, ed. Zuhayr Mikdashi, Sherrill Cleland and Ian Seymour (The Middle East Research and Publishing Centre, Beirut).

Oilfields of the world, E. N. Tiratsoo (Scientific Press Ltd).

Energy: From Surplus to Scarcity?, ed. K. A. D. Inglis (Applied Science Publishers Ltd).

The international oil industry has a particularly good trade Press. We have found *The Petroleum Economist*, *The Petroleum Times*, the *Oil and Gas Journal*, the *Middle East Economic Survey* and *Petroleum Intelligence Weekly* most helpful. *The Financial Times*, *Management Today* and *Fortune* have also been useful to us during our researches. In addition to these sources we have drawn heavily on material in the publications of the Organization of Petroleum Exporting Countries (OPEC), and the oil companies' house journals.

Index

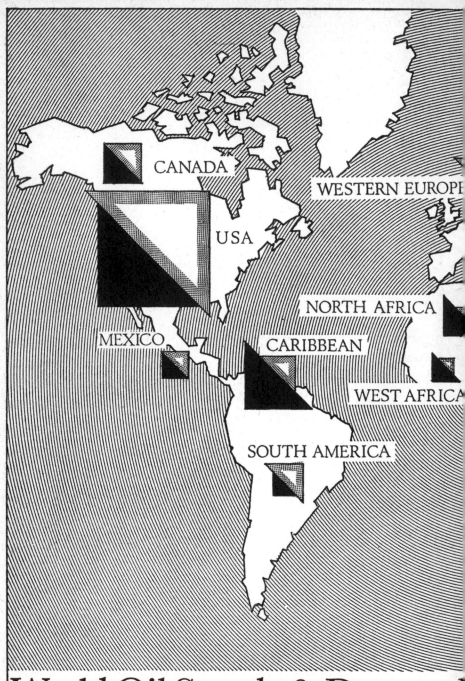

CANADA

WESTERN EUROPE

USA

NORTH AFRICA

MEXICO

CARIBBEAN

WEST AFRICA

SOUTH AMERICA

World Oil Supply & Demand